Superhydrophobic Metal Surfaces

David J. Fisher

Copyright © 2024 by the authors

Published by **Materials Research Forum LLC**
Millersville, PA 17551, USA

All rights reserved. No part of the contents of this book may be reproduced or transmitted in any form or by any means without the written permission of the publisher.

Published as part of the book series
Materials Research Foundations
Volume 167 (2024)
ISSN 2471-8890 (Print)
ISSN 2471-8904 (Online)

Print ISBN 978-1-64490-316-2
ePDF ISBN 978-1-64490-317-9

This book contains information obtained from authentic and highly regarded sources. Reasonable efforts have been made to publish reliable data and information, but the authors and publisher cannot assume responsibility for the validity of all materials or the consequences of their use. The authors and publishers have attempted to trace the copyright holders of all material reproduced in this publication and apologize to copyright holders if permission to publish in this form has not been obtained. If any copyright material has not been acknowledged, please write and let us know so we may rectify in any future reprint.

Distributed worldwide by

Materials Research Forum LLC
105 Springdale Lane
Millersville, PA 17551
USA
http://www.mrforum.com

Printed in the United States of America
10 9 8 7 6 5 4 3 2 1

Table of Contents

Introduction 1
Aluminium 6
    AA1050 15
    AA1060 16
    AA2024 16
    AA3003 17
    AA5051 18
    AA5052 18
    AA5082 19
    AA5083 20
    AA5085 21
    AA6061 22
    AA6063 27
    AA6082 28
    AA7075 29
    AMG 29
    AMS4037 29
Cobalt 29
Copper 30
Iron 45
    AISI1018 56
    C45 56
    Q235 57
    GCr15 60
    17-4 64
    AISI304 65
    AISI316 65
    X90 65
Magnesium 66
    AZ31 66

| AZ31B 71
| AZ61 72
| AZ91 74
| AZ91D 76
| MA8 77
| ZK60 78
Nickel 80
Titanium 80
| Ti-6Al-4V 85
Tungsten 90
About the Author 93
References 94

Introduction

Superhydrophobicity is generally defined as being that property of a surface in which water droplets take up an apparent contact-angle greater than 150°, together with a contact-angle hysteresis of less than 10° and a sliding-angle of less than 5°. It is a property which occurs in Nature, and a great deal of effort has been aimed at mimetically exploiting the superhydrophobicity exhibited, for example, by the lotus leaf.

The fact that the droplets are very far from wetting the surface upon which they stand leads on to many associated tendencies, such as impeding fogging, icing and corrosion. The aim of the present work is to cover the ways in which superhydrophobicity has been imparted to metals. Metals themselves tend more naturally to be hydrophilic and so imparting superhydrophobicity relies upon adding some sort of coating.

William Barthlott (1946-), the discoverer of the lotus effect, noticed that two factors were involved in producing the effect. One was the presence of a waxy material, and the other was the existence of numerous microscopic bumps. The hydrophobic nature of the waxy material causes the water-drops to minimize their contact area with the surface by increasing the angle between the water surface and the leaf surface. The bumps then push the behaviour to superhydrophobic levels, as air which is trapped between the bumps increases the contact-angle. Because of the greater contact-angle, the drop becomes almost spherical in shape and rolls off the leaf.

Based upon the contact-angle and the associated wetting behavior, solid surfaces can be grouped into 4 classes: superhydrophilic (contact-angle less than 10°), hydrophilic (contact-angle between 10° and 90°), hydrophobic (contact-angle between 90° and 150°) and superhydrophobic (contact-angle greater than 150°). The contact-angle and its hysteresis are the most important parameters involved in judging the degree of superhydrophobicity of a solid surface. The contact-angle is used as a common measure of wettability. Contact-angle hysteresis is a measure of the stickiness of the water drop to the solid surface, and is the difference between the advancing angle and the receding angle. If drops are to roll off very easily, a high contact-angle and a low hysteresis are essential. The contact-angle of a typical hydrophobic solid surface is between 100° and 120° and can be increased by increasing the surface roughness.

Superhydrophobicity is often explained by invoking the apparently unrelated Leidenfrost effect, in which water drops on a hot solid surface run around freely. This is because a film of vapour is formed between the surface and the drop, and the latter never makes contact with the solid surface. The Leidenfrost effect is possible if the apparent contact-angle is 180°, hence the analogy with superhydrophobicity. The capillary length of water is about 2.7mm under ambient conditions. If the drop is smaller than the capillary length, it is considered to be small and, in that case, the gravitational force can be neglected and surface tension predominates. When a small drop is placed on an ideal horizontal solid surface, three interfaces exist: solid/liquid, liquid/vapour and vapour/solid.

Overall, the creation of superhydrophobic surfaces requires a combination of surface roughness creation and surface free energy reduction. The wettability of a solid surface is determined both by the microscopic geometry and by the chemical composition of the surface. Liquid droplets contact the substrate so as to form a solid-liquid-gas contact line. When the droplet reaches a steady state on the solid surface, it takes up a certain angle with respect to the surface. At the intersection of the solid-liquid-gas 3-phase contact, the angle between the tangent plane of the liquid-gas boundary and the solid-liquid boundary is defined as the water contact-angle of the droplet on the surface. The value of this angle is an important means for assessing the wettability of solid surfaces. Each has its own associated interfacial surface tension. Thomas Young, of course, established the basic shape of a liquid drop on a solid. The contact-angle for a completely hydrophobic spherical drop is 180° and the angle for a completely hydrophilic wetted surface is 0°. If the surface tensions can be assumed to be material constants, the contact-angle is also a material constant. Young's equation was however derived for smooth surfaces, and not for rough surfaces. The relationship between contact-angle and surface tension was originally studied by Yang, Wenzel and Cassie. Wenzel assumed that the liquid completely wetted a rough surface whereas Cassie and Baxter assumed that a drop simply sat on a rough surface, with air trapped beneath it. The Wenzel model is usually applied to hydrophilic surfaces while the Cassie & Baxter model is more often used to describe the contact-angle of hydrophobic or superhydrophobic surfaces. In the more developed Pillar model the contact-angle depends upon the fractal dimension and the upper and lower limit-lengths of the fractals, while using Wenzel/Cassie states as the basis of analyses. Surfaces with high contact-angle and low sliding-angle have better self-cleaning effects. There are differing views concerning contact-angle hysteresis. Some relate friction at the triple-phase interface to hysteresis of the contact-angle.

The Young equation was modified by Wenzel so as to describe the relationship between roughness and contact-angle. This was done by introducing a so-called roughness factor: the ratio of the actual surface area to the geometrical surface area. The roughness factor is equal to unity for smooth surfaces, and Young's equation is then retrieved. The term, contact-angle, is correct only for smooth surfaces. The term, apparent contact-angle, is applicable to rough surfaces. The Wenzel theory argues that the solid/liquid and solid/vapour interfacial surface tensions are somewhat increased by the increased rough surface area while the liquid/air interfacial surface tension is unchanged. A greater contact-angle is therefore taken up in order to balance the increased surface tension. Even if the smooth solid surface is hydrophilic, the roughened solid surface can become superhydrophilic. If the smooth surface is hydrophobic, the roughened surface can become superhydrophobic. The Wenzel theory was further extended by Cassie and Baxter for the case of porous heterogeneous surfaces. When the surface roughness is higher, it is not essential that the liquid should fill the entire solid surface. The liquid need contact only the peaks of the surface and not enter the valleys. While the bulk of the liquid contacts the peaks of the surface constituting the solid/liquid interface, the remainder contacts any vapour present in the valleys of the solid surface constituting the liquid/vapour interface. Superhydrophobicity is imparted by a suitable choice of roughness, surface texture and added low surface-energy materials.

As noted, there are two main forms of wetting of a rough solid surface: Wenzel and Cassie-Baxter, the modifications of the Young model for smooth surfaces. The Wenzel model assumes that water droplets will fully infiltrate rough surface structures. The Cassie-Baxter model assumes that the contact-angle will increase when there are microstructures on the surface because air is trapped in any gaps so as to form an air cushion which supports the water droplets. When the solid surface is hydrophobic, the static contact-angle alone is not adequate for fully describing the wettability. It is essential to measure the dynamic contact-angle. As the volume of a droplet is increased, the contact-angle also increases and the contact boundary of the solid/liquid interface tends to advance. Above a threshold size, the contact boundary of the solid-liquid interface moves outwards, with the corresponding contact-angle being termed the advancing contact-angle. As the volume of a droplet is decreased, the contact boundary of the solid/liquid interface tends to recede. When the volume of the droplet decreases to a certain threshold, the contact boundary of the solid/liquid interface of the droplet moves inwards, with the corresponding contact-angle being termed the receding contact-angle. The contact-angle hysteresis is then the difference between the advancing contact-angle and the receding contact-angle. The smaller the contact-angle hysteresis, the easier it is for the liquid to leave the surface.

The various methods used to obtain superhydrophobic surfaces can be divided into two main types: top-down and bottom-up. In the former case, the required surface is obtained by etching. In the latter case, the same effect is obtained by chemical deposition. A combination of the two methods can also be used. If the treated surface is not then superhydrophobic, post-treatment with hydrophobic materials can be used to obtain superhydrophobicity. Common hydrophobic materials include silanes, and fluorinated or hydrocarbon thiols. The inclusion of micro-fibres or nano-fibres can provide a good surface texture for water-repellent surfaces.

If a surface can be created with a very low fraction of air at the nano-scale, superhydrophobic surfaces can be produced even from hydrophilic materials. The bottom-up approach is good for preparing very low air fraction superhydrophobic surfaces. The bottom-up methods include chemical vapour deposition, electrochemical deposition and layer-by-layer deposition. Electrochemical processes are capable of controlling both the surface roughness and the surface morphology. Common electrochemical processes are anodization, electrodeposition of conductive polymers, electrodeposition of metals and metal oxides, and electroless galvanic deposition. The anodization creates a porous nanostructured oxide layer. In electroless galvanic deposition, spontaneous deposition of metallic ions occurs when they contact a metallic surface with a lower oxidation potential. The advantages of the bottom-up and top-down approaches can be combined to produce superhydrophobic surfaces with a two-scale roughness. A micro-scale rough surface is first produced by using the top-down approach, and nano-scale roughness is then added by using the bottom-up approach.

Anti-icing is a case in which superhydrophobicity can be effective. Ice adhesion can impair the aerodynamics of aircraft. The superhydrophobic properties of surfaces weaken the adhesion between ice and surface, leading to easy removal by normal or shear forces. A higher contact-angle and a low contact-angle hysteresis are responsible for normal and shear forces, respectively. Icephobic surfaces have a shear strength of between 150 and 500kPa. Superhydrophobic surfaces are sometimes ineffective as anti-icing materials. Ice-accretion on the surfaces can be delayed, when compared with that of flat hydrophobic surfaces, but gradually damages the surface microstructure during icing and de-icing, thus reducing the anti-icing properties. Adhesion-strength increases however, in a humid environment at low temperatures, due to condensation and an associated anchoring effect. The superhydrophobicity prevents icing by forcing water to merge into large drops which then roll off a surface before freezing can occur. The formation of ice can be quite complicated and may involve the formation of quite exotic shapes[1].

True self-cleaning surfaces are those which combine superhydrophilicity and photocatalysis to break down dirt and wash it away. Superhydrophobic surfaces are extremely dry, and repel water drops. So these surfaces do not in fact clean themselves but, when water drops roll over the surface, they wash away dirt. Raindrops should fall at high speed in order that dust particles be washed away. Corrosion resistance, and drag-reduction in underwater applications, are other beneficial effects of superhydrophobicity. When a superhydrophobic surface is completely immersed in water the entrapped air is separated from the moving water. If the air pockets cover a sufficiently large area, the superhydrophobicity can reduce skin friction and cause a slip effect. The degree of superhydrophobicity is decreased if the amount of entrapped air is reduced. It is therefore critical to maintain air pockets in underwater applications.

Biological fouling, also known as biofouling, is the accumulation of the biological matter on a solid surface, together with deposits of corrosion, ice and suspended particles. If a superhydrophobic surface present, it reduces the contact area between water and a solid surface, thus restricting the amount of biological matter reaching the surface. Avoidance of biofouling is possible only if air pockets can be stabilized within pores so as to prevent biological matter from adhering to the solid surface. Since the volume of water is so large, biofouling cannot be characterized by an apparent contact-angle, contact-angle hysteresis or sliding angle. The fraction of wetted area is the only available measure of biofouling superhydrophobicity. Self-propulsion of droplets can occur on a superhydrophobic plant surface such as a lotus leaf. Self-propulsion releases the droplet from the surface/air interface, and any associated adhesion, and leaves it exposed to any forces that can perhaps transport the droplet over significant distances. Gravity and air currents can lead to total removal of the propelled droplets. This can also remove contaminant particles from the surface if they are similar in size to the water droplets[2].

A quite spectacular phenomenon can occur upon letting a water droplet fall onto a pore in a superhydrophobic plate on a water surface. There can be a spontaneous transformation of surface energy into gravitational potential energy[3]. On the basis of this self-capturing phenomenon, a power-free water-pump was created[4] which comprised a superhydrophobic plate with a pore, mounted on a leak-proof cylindrical container filled with water. This led to the anti-gravity long-distance transport of water. The use of a superhydrophobic surface, having the ability to withstand high pressures and exhibiting low adhesion, could constitute a power-free pump. The lifting-height of the pump was of the order of 100mm, and increased with decreasing pore diameter. The transport capacity of the pump was unaffected by the tilt of the pump body or by the tube diameter. Continuous delivery of water was possible over distances of the order of metres.

The oxide films on aluminium surfaces easily absorb oil and other pollutants, causing damage to the surfaces and reducing the service life of the material. A superhydrophobic surface offers self-cleaning, anti-icing, anti-corrosion and anti-fouling properties, but suffers from poor mechanical durability. It is therefore important to improve that durability. Increasing the substrate roughness is one of the available means for improving the durability of a superhydrophobic surface, in that an increase in the substrate roughness changes the microstructure and thus influences the mechanical behaviour. The roughness can be increased by laser-processing, chemical etching and sanding. Sandblasting is widely used to clean surfaces and create rough surfaces, but its use to create a rough substrate for superhydrophobic surfaces is still being explored.

The methods which are used to produce a superhydrophobic metal surface can also be divided into direct and indirect. The latter involves the creation of a superhydrophobic coating on the metal surface. Coatings can greatly improve the superhydrophobicity, corrosion resistance and other properties of metal surfaces but they cannot change the wettability of the metal itself. Poor adhesion between a coating and a metal substrate also limits their service life. The direct method instead involves a process which endows the metal surface itself with superhydrophobicity, although this may lead to considerable loss of material.

Aluminium

The creation of low surface-energy non-fluorinated dual-scale so-called *Allium giganteum* structured superhydrophobic surfaces on 5N-purity aluminium was carried out using a 1-step electrodeposition method[5]. The maximum contact-angle of 168.6° was found for a deposition voltage of 30V. Air-exposure and 3.5wt%NaCl immersion tests indicated a long-lasting surface life.

Ultra-fast (<120s) 1-step electrodeposition was used[6] to create a fluorine-free superhydrophobic 5N-purity aluminium surface which consisted of a manganese palmitate complex with a dual-scale hierarchical papillae structure having a low surface energy. The as-prepared surfaces exhibited extremely low surface-adhesion and an excellent self-cleaning capability. There was a marked augmentation of the charge-transfer resistance of corrosion and a concomitant anti-corrosion behaviour, with a corrosion-inhibition efficiency of up to 99.94%.

Mirror-finish superhydrophobic 4N-purity aluminium surfaces were prepared[7] via the formation of anodic alumina nanofibers and subsequent modification with self-assembled monolayers. The high-density nanofibers were formed on the surface by anodizing in pyrophosphoric acid solution. The nanofibers became entangled and bundled by further anodizing at low temperatures, and the aluminium surface was completely covered in long floppy nanofibers. The surface had a contact-angle of less than 10°. When the nanofibre-covered surface was modified with n-alkylphosphonic acid self-assembled monolayers, the water contact-angle suddenly shifted to superhydrophobic with a contact-angle greater than 150°. The angle increased with applied voltage during pyrophosphoric acid anodizing, anodizing time and number of carbon atoms in the monolayer molecules on the alumina nanofibers. By optimizing the anodizing and modification conditions, superhydrophobic behaviour could be obtained by using a pyrophosphoric acid anodizing period of only 180s and subsequent immersion. The superhydrophobic aluminium surface also exhibited high (99%) reflectance across most of the visible spectrum.

An alternative method[8] for enhancing the superhydrophobicity of 3N-purity aluminium surfaces led to a contact-angle of about 153° due to the formation of an hierarchical structure via the grinding and polishing of micro-grooves into the surface, combined with simultaneous exposure to Hydrochloric acid and dodecanoic acid. There was a reaction between the fatty acid and the aluminium surface. The metal and its oxide appeared to be involved, and free-aluminium was anchored to fatty-acid molecules and to alumina molecules in the medium. Both metallic aluminium and its oxides were presumably required to form the compound that was responsible for the superhydrophobicity. An important factor is the existence of features which hold air and prevent water droplets from coming into contact with the solid surface. The homogenous Wenzel model predicts that a water droplet can penetrate asperities, while the composite Cassie–Baxter model suggests that a droplet can be suspended above asperities when a gas is trapped in the cavities of a rough surface.

A superhydrophobic coating was created[9] on a 3N-purity aluminium surface by anodization in a sulphuric acid electrolyte, followed by surface modification with myristic acid. The contact-angle of the coatings increased from 114.1° to 155.2° upon increasing the anodization voltage from 0 to 22V. The contact-angle markedly improved when the anodization voltage reached 20V. When the voltage was further increased to 22V, the contact-angle and sliding-angle worsened from 155.2° and 3.5 to 152.8 and 7.0 , respectively. The as-prepared coating had an hierarchical micro-nano structure, with a static water contact-angle of 155.2° and a sliding-angle of 3.5 . It retained a contact-angle of up to 151.1° following sand-blasting for 60 s and remained stable after exposure to acidic and alkaline solutions. Following exposure to ultra-violet and water condensation cycles for 7 days, the coating remained superhydrophobic. The coating also exhibited excellent self-cleaning and anti-icing capabilities; ice-adhesion strengths as low as 0.065MPa were measured. There was a large reduction in the corrosion current-density, and the protection efficiency of the as-prepared coating attained 99.75%.

Superhydrophobic 3N-purity aluminium with an hierarchical micro-nano structure was produced[10] by combining anodizing, chemical etching and surface-modification using SiO_2-polydimethylsiloxane coating via chemical vapour deposition. The anodizing produced a cylindrical nanopore (70 to 90nm) network. During etching, the diameters of the pores increased, and the surface modification deposited SiO_2-polydimethylsiloxane nanoparticles over them. Water contact-angle measurements showed that samples of anodized and anodized-etched aluminium, following modification, became superhydrophobic with values of 153° and 154°, respectively, and a sliding-angle of less than 1°. The charge-transfer resistance of the superhydrophobic surface of modified anodized-etched material was 16700kΩcm^2; some 900 times higher than that of pure aluminium. There was also a positive shift of 0.385V in corrosion potential and a reduction of more than 3 orders-of-magnitude in the corrosion current.

Contact- and sliding-angles on 2N purity aluminium were measured[11] from -10 to 30C in relative humidities of 30, 60 or 90%. Rough surfaces were treated with perfluoro-alkyltri-ethoxysilane [$CF_3(CF_2)_nCH_2CH_2Si(OC_2H_5)_3$, n = 6 to 8] polymer, palmitic acid [$CH_3(CH_2)_{14}COOH$] and silicon rubber [$(CH_3O)_2CH_3SiO[(CH_3)_2SiO]$] so as to produce superhydrophobic materials. Calculations of the solid/liquid contact-area fraction quantitatively explained an increased wettability which was characterized by a decreasing contact-angle and an increasing sliding-angle under low-temperature high-humidity conditions and indicated a transition from Cassie-Baxter to Wenzel behaviour on rough surfaces. A loss of superhydrophobicity during condensation could be completely restored by drying at room temperature.

The hierarchical growth of boehmite (γ-AlOOH) film on 2N5-purity aluminium foil was achieved[12] by using a solution-phase synthesis route. The resultant film comprised 3-dimensional micro-protrusions which were assembled from well-aligned nano-needles; a biomimetic version of lotus leaves. The surface following hydrophobization had a water contact-angle of 169° and a sliding-angle of about 4° for 5μl droplets. This was attributed to the combined effects of dual-scale roughness at the micro- and nano-meter level and the low surface energy of stearic acid. The film exhibited relatively good adhesion to the aluminium substrate and retained superhydrophobicity after ultrasonic treatment or long-term storage. There was a partial loss of superhydrophobic capability following abrasion.

Chemical etching was used[13] to prepare superhydrophobic aluminium surfaces having a water-contact angle of 154.8° and a sliding-angle of about 5°. The etched surfaces exhibited irregular micro-scale plateaux and caves within which there were nano-scale block-like convex and hollow features. The superhydrophobicity existed only for some structures in which the plateaux and caves were appropriately ordered. The resultant surfaces exhibited good self-cleaning properties. The results indicated that it was possible to produce a superhydrophobic surface on a hydrophilic substrate by suitably adjusting the surface structure so as to provide more spaces which trapped air.

Superhydrophobic surfaces were created[14] on 100μm pure aluminium by means of 1-step nanosecond laser-processing. The thin sheets were micro-patterned using 500mW ultraviolet laser pulses and a direct laser writing technique. The microstructure exhibited blind micro-holes which improved the interface between water, air and solid, and thus enhanced the wetting of the surface. The geometrical changes were related to chemical changes at the surface which improved the degree of hydrophobicity. The laser-processed micro-holes exhibited near-superhydrophobic behaviour, with a static contact-angle of 148°.

Anodic oxidation and self-assembly processes were used[15] to prepare superhydrophobic aluminium alloy surfaces having a water-contact angle of 157.5° and a sliding-angle of 3°. These resulted from its hierarchical micro-nano structure and the assembly of low surface-energy fluorinated components upon it. The untreated alloy substrate was hydrophilic, with a contact-angle of about 97.9°. After some modification, but without anodic oxidation, the surface was hydrophilic with a contact-angle of about 114.7°. Water droplets spread completely on surfaces which were modified only by anodic oxidation. The essentially zero contact-angle suggested that the anodized alloy surface was superhydrophilic. This was because the volume of the hydrophilic alumina and the roughness of the surface greatly increased during anodizing. A transition from superhydrophilicity to superhydrophobicity was possible by adjusting the modification process for the surface. The superhydrophobic surface maintained its nature following abrasion with P400 grit silicon carbide paper over 0.4m, and P800 grit sandpaper over 0.8m, under an applied pressure of 3.60kPa. The surface offered an excellent corrosion resistance and self-cleaning capability together with long-term stability.

Superhydrophobic surfaces were produced on aluminium by shot-peening, dislocation-etching and immersion in solutions of stearic, myristic and decanoic acids[16]. The dislocation-etching produced an hierarchical structure at the nano-scale and micro-scale. Surfaces which were etched with stearic acid had a low surface energy, with a contact-angle of about 157° and low adhesion for 8μl drops. Increasing the etching time increased the static contact-angle, as it increased the roughness and the number of micro-pores increased. According to the Cassie-Baxter law, the hierarchical structure trapped air-bubbles and decreased the area of contact-surface with water. The fatty acids also increased the corrosion resistance of the surface, and electrochemical impedance tests revealed a 25-fold improvement, with the corrosion-current density decreasing by about one decade.

Superhydrophobic aluminium surfaces with controllable adhesion were prepared by means of femtosecond laser ablation, and that adhesion could be modified from extremely low to high by adjusting laser-processing parameters[17]. Various hierarchical structures comprising micro/nano-scale features could be produced simply by adjusting the processing parameters so as to lead to a range of wetting abilities. Cleaned samples were irradiated with linearly-polarized light with a wavelength of 1030nm, using a repetition-rate of 75kHz and a pulse-width of 1000fs. The output-power ranged from 1000 to 10000mW. The resultant changes in surface morphology depended sensitively upon the scanning-speed, laser-power and scanning-interval and could cause a transition from Cassie to Wenzel behaviour.

A coating was prepared[18] which consisted of an upper hydrophobic layer of low surface-energy hexamethyldisilazane-modified nano-silica and an intermediate connecting layer of hydrolyzed glycidoxypropyltrimethoxysilane that was covalently bonded to the superhydrophobic surface and a hydrophilic aluminium substrate. The superhydrophobic coating had a multi-scale fractal morphology with a fractal dimension of 2.2, and asperities on a length-scale of 32 and 630nm. The maximum water contact-angle of the surface was 170, with a sliding-angle of about 1°. The surface exhibited a good self-cleaning capability for both hydrophobic and hydrophilic contaminants. There was less than 5%-detachment of the coating following cross-hatch adhesion tests, and superhydrophobicity was maintained for up to 150cm in abrasion tests. A corrosion-inhibition efficiency of about 99% was also exhibited by the surface, and there was poor adhesion to gram-negative and gram-positive bacteria.

Laser-texturing was used to create[19] a regular dimple-pattern on aluminium surfaces which, following stearic-acid treatment, became superhydrophobic. Five types of regular dimple-pattern arrays were produced on the surface, with diameters of 0.2, 0.4, 0.6, 0.8 or 1.0mm. The wettability of the laser-textured samples could be regulated by choosing the dimple dimensions during laser-processing. When the diameters of the dimples were increased, the water contact-angle decreased and the water sliding-angle increased. Fluorescence methods could identify those zones on the surface which were penetrated by sufficiently small molecules and, considering Cassie-Baxter theoretical calculations, the fluorescence method revealed the air-trapping capability of the surface. The materials also exhibited a marked mechanical stability, when compared with bare aluminium, because of the occurrence of case-hardening of the laser-patterned surfaces.

Aluminium foil with a roughened surface was first prepared by anodic treatment in neutral aqueous solution[20], and pitting corrosion by NaCl. The surface of untreated foil was smooth but became coarse, with a labyrinthine hole and step topology following being anodizing. The original hydrophobic surface, with a contact-angle of about 79°, became superhydrophilic, with a contact-angle of less than 5°, following anodizing. The superhydrophilic surface then became superhydrophobic, with a contact-angle greater than 150°, following modification with oleic acid. The icing of the various foils was investigated at -12C. The mean total times required to freeze a 6µl water droplet on the superhydrophilic, untreated and superhydrophobic foils were 17s, 158s and 1604s, respectively; the superhydrophilic surface accelerated icing while the superhydrophobic surface delayed icing. This transition was attributed to the differences in the contact areas of the water droplet with the substrate, because an increase in contact-area will accelerate heat transfer and thus icing. A decrease in contact-area will delay heat-transfer and hence icing.

A superhydrophobic surface was created on an aluminium substrate by using a sol–gel method which involved immersing the clean pure substrate into a solution of zinc nitrate hexahydrate and hexamethylene tetra-amine in various molar ratios and constant 0.04mol/l total concentration[21]. After heating (95C, 1.5h), it was modified with alkane thiols or stearic acid. When the above molar ratio was changed from 10:1 to 1:1, the contact-angle was greater than 150°. The best surface had a water contact-angle of about 154.8°, with an angle-hysteresis of about 3°. The surface of the films was composed of ZnO and Zn-Al layered double hydroxide in flower-shaped arrangements. The flower-like porous structure and the low surface energy led to the superhydrophobicity.

Superhydrophobic surfaces were prepared by spray-coating[22]. A static water contact-angle of about 154° was obtained by depositing stearic acid on an aluminium alloy, and the coating exhibited a high (circa 30°) contact-angle hysteresis. Superhydrophobic surfaces with a static contact-angle of about 162° and 158°, and a contact-angle hysteresis of about 3° and 5°, respectively, were obtained by incorporating nanoparticles of SiO_2 and $CaCO_3$ into stearic acid. The excellent hydrophobicity was attributed to the synergistic effects of micro/nano-roughness and low surface energy. Study of the wettability from 20 to -10C showed that the superhydrophobic surface became hydrophobic at low temperatures.

An homogeneously structured superhydrophobic surface with a gradient non-wettability was created[23] by combining chemical etching and vapour-diffusion modification. The as-prepared surface exhibited a marked gradient of water repellency, with the water contact-angle being between 162 and 149°. The sliding-angle exhibited a corresponding change from 3 to 11°. The gradient nature of the non-wettability led to a droplet surface-adhesion that ranged from 19μN at the most hydrophobic end to 57μN at the other end. Because of the difference in water-adhesion force, droplets tended to roll well in a specific direction; the gradient non-wettability.

The directional caterpillar-like rolling of droplets along the ridges of inclined ratchet-like superhydrophobic surfaces was observed[24]. The superhydrophobic coating comprised micron-scale particles and 50 to 100nm nano-holes. In the opposite direction, the droplet movement depended only upon the end of triple-phase contact-lines while the front of the contact line was pinned. Sliding-angle measurements indicated that the ratchet-like superhydrophobic surfaces exhibited a directional drop-retention behaviour. A reduction in the rise-angle, in the height of the ratchet's ridge and in the volume of the droplet could markedly increase the directional difference in droplet retention on the ratchet-like surfaces. It was concluded that superhydrophobicity and periodic ratchet-like microstructures were pivotal to directional droplet-sliding in the 1-dimensional case.

Table 1. Effect of HCl-etching time on aluminium surfaces

HCl (min)	Modification	Contact-Angle (°)	Rolling/Sliding-Angle (°)
0	a	74.7	-
4	a	25.6	-
4	a	169.2	4.2
2	a	107.7	135.7
3	a	154.3	21.9
6	a	168.0	5.1
8	a	166.6	3.8
4	b	155.9	45.7
4	c	164.8	6.8

a) 10min modification with dodecanethiol-myristic mixture, b) 10min modification with dodecanethiol, c) 10min modification with myristic acid

Common etchants and organic acid reagents were used[25] to explore the effects of etching and composition modification in producing superhydrophobic aluminium surfaces. An unstable layer was found following HCl-etching, which consisted of $AlCl_3/Al_2O_3$ particles surrounded by Al_2O_3 plates. This layer could be washed off. Treatment with dodecane thiol and myristic acid also modified the microstructure via a mild etching mechanism. A synergistic effect resulted from using their mixtures, leading to the finest grains and highest number of carbon chains on the final surface. Flexible superhydrophobic aluminium sheet was used to subject droplets to rolling and coalescence. This revealed the conditions which yielded the best superhydrophobic surfaces; those with the highest water contact-angles and lowest sliding-angles, as well as the shortest contact-time during droplet bouncing. The best water-repellent properties, as indicated by the contact- and rolling-angles (table 1), were obtained by HCl-etching for 240s and treatment with mixed solutions. A pristine aluminium was usually smooth at the micrometre scale, with just a few scratches and patch-like defects. Following HCl-etching, the water contact-angle had decreased from about 76° to about 26°. This was attributed to Wenzel-type contact, in which surface-roughness increased wettability. Mixed organic solution not only supplied the surface with hydrophobic groups, but also modified the surface geometry transformed plates into small pieces so as to produce a surface with an homogeneous micro-nano structure. This was because, as noted above, the modification treatment was also an etching process.

Superhydrophobic aluminium surfaces were investigated[26] as a means for encouraging drop-wise condensation. The required superhydrophobicity was obtained by etching, and then depositing fluorosilane so as to lower the surface energy. Some samples were etched by immersion in aqueous $FeCl_3$ solution. Other samples were immersed in aqueous $CuCl_2$ solution. Experimental tests on pure steam condensation (saturation-temperature of about 103C, inlet cooling-water temperature of about 25C and cooling-mass flow-rate of about 0.11kg/s) were performed on the resultant materials. A heat-transfer coefficient greater than $50kW/m^2K$ was measured: some 4 times greater than that for hydrophilic untreated aluminium. Moreover, the different etchants nano-textured the aluminium surface with differing surface morphologies. Both etchants led to contact-angles of about 156° and contact-angle hystereses lower than 10° at room temperature (table 2). Both substrates promoted drop-wise condensation. In the saturated-vapour environment, condensate droplets grew and moved on the surfaces in the Wenzel state.

Table 2. Contact-angles of etched aluminium surfaces

Etchant	Advancing Contact-Angle (°)	Receding Contact-Angle (°)	Hysteresis (°)]
none	85	17	67
$FeCl_3$	155	149	7
$CuCl_2$	158	149	9

The possibility of restricting the size of condensate droplets on superhydrophobic materials showed[27] that these materials offer the ability to prevent dripping from cooled ceiling panels. Superhydrophobic and plain aluminium sheets were tested in a relative humidity of 80% and a dew-point of 21.4C. The sheets were cooled at 6C for 8h. The temperature of the sample surfaces was about 13.5C during testing. After 8 hours of condensation, the diameter of the droplets on the superhydrophobic sheet was less than 150μm, while the largest droplets on the plain sheet were about 4mm. The size of the largest droplet on superhydrophobic aluminium was usually less than 100μm. During most of the tests, the largest droplet which appeared after 5.5h was 151μm; below the normal visual acuity limit of 300μm. Droplets grew continuously on the plain aluminium sheet and the largest (4mm) drop was found at the end of the test.

Cyclic chemical etching was used[28] to create superhydrophobic surfaces with a honeycomb structure. Samples which were etched 8 times exhibited micro-nano scale honeycomb cavities and had a water contact-angle of 135°. When further treated with octadecane thiol methanol solution, the contact-angle was 153.1°. The contact-angle of the unetched surface was about 46.1°, making it hydrophilic due to the inability of the smooth surface to capture air. The angle increased rapidly with increasing etching time. The roughness increased from 117nm to 779nm with increasing etching time, and the change in surface morphology from smooth to honeycomb pores resulted in an increase in the amount of air which was captured. Modification with the octadecane thiol covered the etched surface with a monomolecular film which was formed by adsorption and reaction between the octadecane thiol and the etched surface, thus reducing the surface free-energy and producing a superhydrophobic surface. The octadecane thiol was physically absorbed, and then the S-H bond in the CH_3-$(CH_2)_{17}$-SH molecule was broken and a strong sulphur-aluminium bond was formed. The outermost part of the monomolecular film was composed of CH_3, and this could greatly reduce the surface free-energy of the monolayer. The corrosion-current densities of the original, etched and superhydrophobic materials were 0.000816, 0.00166 and 0.000748A/cm^2, respectively. The corrosion-current density of superhydrophobic material was lower than that of the original and etched materials. The corresponding corrosion-voltages were -1.17, -1.21 and -1.26V. The combination of the honeycomb microstructure and the low surface energy monolayer formed a so-called air-film layer that could block corrosive ions from direct contact with the aluminium surface. Solutions of NaCl could easily spread on etched surfaces lacking surface-energy modification and lead to corrosion. Chalk-dust was used to mimic pollution. Water droplets rolled on the superhydrophobic surface and washed away the chalk.

Aluminium surfaces were treated electrochemically, and the addition of hydrochloric acid to the electrolyte had an appreciable effect upon the wettability[29]. The static water contact-angles were 115.6° and 129.2° when the electrodeposition times were 5 min and 10 min, respectively. Surfaces which were prepared by using 2V for 20min had a static water contact-angle of 155.8° when 0.015M HCl was added. The static contact-angle remained above 150° following exposure to 3.5wt%NaCl solution for 30 days. The low contact-angles for short electrodeposition times were attributed to insufficient surface etching of the aluminium. The corrosion-resistance of as-prepared superhydrophobic aluminium surfaces was determined using electrochemical impedance spectroscopy and electrochemical noise techniques. The average corrosion-rate and pitting-intensity were marked inhibited by the as-prepared superhydrophobic surface. Just scratches originating from polishing were present on plain samples, while the superhydrophobic surface was much rougher and was covered with leaf-like sheets of various sizes. Compact micro- and nano-sheets which were stuck to the surface could harbour considerable amounts of air and thus exhibit superhydrophobicity. This had a marked effect upon the corrosion-resistance in 3.5wt%NaCl solution.

A superhydrophobic surface was prepared on an aluminium substrate, with anodization and low-temperature plasma treatment being used to create a micro–nano structure and trichloro-octadecyl-silane being used to modify the roughened surface[30]. The resultant static water-contact angle of the surface was 152.1°. A rougher surface, with micro-nano pores and other features, was produced when a low-temperature plasma treatment was applied to anodized aluminium film, resulting in a static water-contact angle of up to 157.8°.

Nanoparticles of ZnO were deposited onto commercial-purity aluminium by simple immersion and ultrasound treatment[31], followed by surface-energy reduction using stearic acid in ethanol solution. The ultrasound led to more stable superhydrophobic aluminium surfaces when compared with simple immersion. Etching in Hydrochloric acid solution was sometimes used before ZnO deposition in order to create more mechanically stable superhydrophobic surfaces. Use of ultrasound also led to a marked decrease in corrosion current-density and to long-term stability improvements when compared with immersion in 3.5%NaCl solution without etching. The ultrasound could create unique conditions for inducing changes that resulted from cavitation phenomena. In heterogeneous environments, the collapse of bubble was asymmetrical and could affect surface morphology via the creation of micro-jets and shock waves. The distinctive topography could have a marked effect upon the wettability and stability of superhydrophobic surfaces. With increasing immersion time, the water contact-angle increased and reached a plateau after 0.25h. The sliding-angle fell to its lowest value at this point.

AA1050

Single-step anodization, with sulphuric acid as electrolyte, was used[32] to produce micro-nano structures on AA1050 substrates. The surface energy of the anodized layer was then reduced by stearic-acid modification. An anodizing current of 0.41A, a sulphuric-acid concentration of 15wt% and an anodizing time of 1.5h were the optimum conditions, and led to a contact-angle of up to 159.2° with a sliding-angle of less than 5°. There were irregular so-called bird's nest structures on the surface instead of the highly-ordered honeycomb structures which were expected following anodizing. Only amorphous structures were present on the surface. The Brunauer–Emmett–Teller specific surface area of the anodized layer was 2.55m^2/g under optimum conditions. The surface had retained its hydrophobicity in air after one week and in de-ionized water after 12 weeks.

AA1060

A coral-like superhydrophobic surface was produced on clean AA1060 foil by etching with 1mol/*l* CuCl$_2$ solution for 8s and using 90C hot-water treatment (50min) to create micro/nano roughness and improve the anti-icing behaviour[33]. The as-prepared surface consisted mainly of aluminium and small amounts of AlO(OH). The ambient-temperature (48% relative humidity) of this surface had a contact-angle of 164.8° and a sliding-angle of less than 1°. The coral-like structure indeed exhibited excellent anti-icing properties, with water droplets remaining unfrozen on the as-prepared surface for more than 110min at -6C. Some 71% of the surface remained free from ice when exposed to so-called glaze-ice for 30min.

Superhydrophobic surfaces were created[34] by first spin-coating epoxy resin, and transferring the dendrite-like structures to an AA1060 substrate which was coated with epoxy resin. The superhydrophobic surface was obtained by modification with stearic acid. This combination of an organic coating and a superhydrophobic surface with a dendritic micro-nano structure more effectively prevented the permeation of corrosive ions. The superhydrophobic surfaces had a water contact-angle and sliding-angle of nearly 154° and 0°, respectively. The anti-corrosion properties of the superhydrophobic surface in 3.5wt%NaCl solution were greatly improved by the epoxy resin and by air pockets trapped in the superhydrophobic surface.

AA2024

Sheets of AA2024 with a surface roughness of about 200nm were irradiated in ambient air with linearly-polarized laser light (1030nm) using a pulse-duration of 900fs and a beam-diameter of 25µm[35]. The specimens were held at 100C for 24h in order to accelerate aging. Various regular micro-grid textures (square, triangular, hexagonal-circle) were created on the surface by scanning the laser beam. The inscribed tracks had a depth of 10µm and a width of 25µm. At a certain percentage of textured surface there was a sudden switch from hydrophobic to superhydrophobic behaviour at a contact-angle of 180°. Before the transition there was a general increase in water contact-angle as the hatch-distance decreased, regardless of droplet-volume. The superhydrophobicity of the textured surface was attributed to air which was trapped in the microstructures and held-up the water droplet so as to impede contact with the surface. As the distance increased, the fraction of hydrophilic aluminium surface in contact with liquid increased and led to a reduction in the contact-angle. During laser-ablation, existing organic molecules are removed. Following aging, a higher concentration of carbon arising from the adsorption of molecules on an aluminium oxide layer, post laser-treatment, is detected and causes superhydrophobicity of textured surfaces. Larger (10µl) droplets had a greater weight than 2µl droplets and this was expected to lead to lower contact-angle. A larger droplet meant however that a larger portion of the textured surface was in contact with the droplet, leading to more support from air cushions and surface chemistry. This compensated for the greater weight of heavier droplets. At lower percentages of textured surface the uncertainty in contact-angle measurement was greater. This was attributed to the portion of textured surface in contact with droplets. The contact-area of a 2µl droplet, with a diameter of about 1.7mm, was commensurate with the size of the grid such that a different location of the droplet resulted in a different enlargement and a very different contact-angle. When the droplet was close to a laser groove, its spreading would stop at the groove and lead to a larger contact-angle. When the contrary was located far from laser grooves, greater spreading was possible and led to a smaller contact-angle. For all of the textures, the denser the lattice the greater was the static contact-angle of droplets. In the case of triangle- and circle-patterns, the threshold of superhydrophobicity occurred at a lower percentages of textured surface. This was probably because, for a given percentage of textured surface, the circle and triangle patterns had a greater total of ablated volume. This in turn was because, in order to produce those structures, the scanning-path overlapped more often and thus deepened the grooves at those intersections and trapped more air to support droplets.

AA3003

Intermetallic phases in AA3003 in the strain-hardened and partially annealed state have a strengthening effect, but promote pitting. In order to improve the corrosion resistance, a superhydrophobic surface was prepared[36] by etching with NaOH and modification with stearic acid. The contact-angle increased from about 150° to about 168° upon increasing the etching-time to 10min. Upon further extending the etching-time to 20min there was a slight decrease to about 164°. The surface consisted of micro-rods and nano-sheets. The intermetallic phases provided firm support for the superhydrophobic surface and avoided the promotion of localized corrosion. The formation mechanism of an etched structure was attributed to the synchronous chemical reaction of intermetallic phases and matrix phases in the etchant. Further modification with stearic acid switched the micro/nano hierarchical surface from hydrophilic to superhydrophobic. The as-prepared surface had a maximum static contact-angle of about 169° and a sliding-angle of about 1°. The corrosion inhibition-efficiency exceeded 99% in 3.5wt%NaCl solution. The pinning direction of intermetallic phases in the superhydrophobic surface affected its properties.

AA5051

A near-superhydrophobic surface was prepared on AA5051 by 20W nanosecond laser texturing[37]. The depth of the micro-valleys was about 17μm. During creation of the textures, nano- and micro-particles were generated and re-deposited along each side of the lines. The height of the micro-wall structure increased the depth of the textures and affected the surface roughness; thus influencing wettability. The initial contact-angle was zero, and increased with heat treatment. Upon heating the textured samples at 150 for 0.5h, the contact-angles reached 96.59° and 100.9° for sinusoidal and grid patterns, respectively. These values increased to 134.88° and 136.97°, respectively, for sinusoidal and grid patterns after 72h of heat-treatment. The patterns thus changed from superhydrophilic to near-superhydrophobic during heat treatment. The latter accelerated organic adsorption on the laser-treated aluminium. The change in wettability was attributed to the roughness and low surface energy.

Table 3. Rolling-angles on superhydrophobic AA5082 surfaces

Asperity-Height (nm)	Real Surface Area (μm^2)	Rolling-Angle (°)
11.26	26.87	>90
15.12	27.85	>90
28.22	30.46	>90
29.48	31.94	25
30.37	28.92	<3
136.5	41.37	>90
172.0	52.33	>90

AA5052

Various surface treatments were applied to AA5052 in order to evaluate their effect upon the obtention of superhydrophobic surfaces exhibiting anti-corrosion properties in saline media[38]. The surfaces were obtained using 3 sequential steps: etching in hydrochloric acid solution, treatment with zinc nitrate in an alkaline medium and reduction of the surface energy by using stearic acid. The initial surface condition affected the superhydrophobic and anti-corrosion properties of the coating. Samples which were initially sanded yielded the best results with regard to contact-angle (154°), sliding angle (5.67°) and efficiency of corrosion resistance (89.82%). After immersion corrosion-testing for 28 days, the coating had lost its superhydrophobic nature and instead exhibited hydrophilic behaviour.

A superhydrophobic film was created on AA5052 via the electrodeposition of a Ni–Co alloy coating, followed by modification with 6-(N-allyl-1,1,2,2-tetrahydro-perfluorodecyl) amino-1,3,5-triazine-2,4-dithiol monosodium[39]. The surface had an hierarchical nano-micro structure with a contact-angle of 167.3° and a sliding-angle of about 1°. The surface offered an excellent superhydrophobicity and self-cleaning capability, with a contact-angle of 160° being retained after atmospheric exposure for 240 days. The surfaces were exposed to the open air, at a temperature of 24 to 26C and a relative humidity of 40 to 50%. The water contact-angle changed from 151.3 to 155.6° over 4 weeks. When the exposure-time was longer than 16 weeks, the contact-angles increased from 155.6 to 160.0°. The coatings also greatly improved the corrosion-resistance in 3.5wt%NaCl solution.

Table 4. Surface-roughness and water contact-angle of AA5083

Surface	Roughness (nm)	Contact-Angle (°)	Hysteresis (°)
as-received	1000	80	–
anodized	78.6	30	–
anodized, 2% (PFOTS)	179	160	9
anodized, 5% (KH-832)	134	170	3

AA5082

Three methods were used to obtain mimetic superhydrophobic surfaces on AA5082: short-term treatment with boiling water, HF/HCl or HNO_3/HCl concentrated solution etching[40]. A thin layer of octadecylsilane was then applied via *in situ* polymerization on the pre-treated surface. Each method led to differing hierarchical nano-micro structures. The water contact-angles ranged from 160° to almost 180°. The best result was that for HF-etched surfaces, with a water contact-angle of more than 175°. All of the surfaces exhibited water contact-angles above 150°, only one had a rolling-angle of less than 5°

(table 3), suggesting a transition from Wenzel to Cassie–Baxter behaviour, where, due to high roughness, air-pockets could be trapped and augment the hydrophobic behaviour. Air which was trapped in the water/solid interface also reduced the contact-area between the substrate and the droplet, thus permitting water rolling at a low tilt-angle. This explanation could not be applied when the rolling-angle was greater than 90°. The surface was here still in the Wenzel state where the water droplet fully penetrates surface grooves and the triple-phase liquid/air/solid contact-line is stable and continuous. The adhesive force is proportional to the Van der Waals force and negative pressures are produced by air sealed in the surface. When a droplet is placed on a superhydrophobic surface, the negative pressure can be neglected. When the surface is turned, the surface/air/water contact-line changes from concave to convex and the volume of sealed-in air increases. The latter expansion can increase the negative pressure and thus increase liquid adhesion.

AA5083

The surface of AA5083 was subjected to anodizing in sulphuric acid, and modification with tri-ethoxyoctyl silane (KH832) or 1H,1H,2H,2H-perfluoro-octyltrichloro silane (PFOTS). The superhydrophobic coatings were characterised by their contact-angles and related hysteresis (table 4). Anodizing reduced surface-roughening while further chemical modification increased the roughness[41]. The anodised surface structure was leaf-like before and after modification. This could increase the possibility of air-trapping and superhydrophobicity. Modification using the two types of silane led to the formation of siloxane and CF_2 groups, respectively, on the surface. The effect of the siloxane groups upon superhydrophobicity was greater than that of CF_2 groups. The observation of very high static water contact-angles and very low contact-angle hysteresis confirmed the formation of a composite interface between the coating and water droplets.

Table 5. Roughness of silane-treated AA5085 surfaces after laser pre-treatment

Power (W)	Roughness (µm)
0	0.85
10	0.92
15	0.99
20	1.91
30	2.42

AA5085

The role played by alumina coatings in varying the wettability of AA5085 by using nanosecond laser ablation was investigated[42]. A superhydrophilic surface having a water contact-angle that was close to 0° was first prepared, and then transformed into a superhydrophobic surface having a water contact-angle of about 151.4° by treatment with non-fluorinated N-octyltri-ethoxysilane. In order to analyze the mechanism of change of the contact-angle, the surface roughness following laser treatment was measured (table 5). The original alloy surface, with a thickness of 15μm, was uneven. After 10W and 15W laser-treatment, the surface morphology remained largely unchanged.

Table 6. Effect of sand-peening parameters on AA6061 contact-angle

t	PS(mesh)	ND(mm)	IT(s)	Contact-Angle (°)
0.6	30	4	60	138
0.6	60	5	90	135
0.6	90	6	120	139
0.6	120	7	150	142
0.65	30	6	150	139
0.65	60	7	120	140
0.65	90	4	90	141
0.65	120	5	60	148
0.7	30	7	90	140
0.7	60	6	60	138
0.7	90	5	150	143
0.7	120	4	120	139
0.75	30	5	120	143
0.75	60	4	150	139
0.75	90	7	60	138
0.75	120	6	90	140

IP: impact-pressure, PS: particle-size, ND: nozzle-diameter, IT: impact-time

An obvious change in surface morphology occurred at 20W. The aluminium oxide coating on the laser-treated path was destroyed during laser-treatment. Parallel grooved structures appeared along the beam path, and micro-nano structures emerged at the bottom of the grooves. Non-decomposed aluminium oxide coating existed only at the peaks of the grooves. Within the grooves, the aluminium oxide was completely decomposed by the ablation. The changes were more marked when the laser power was 30W. The effective coating thickness of the superhydrophilic surface was about 20μm from the peak to the bottom of the groove. Hydrophilic groups, Al-OH(-OH), which were generated by coating-decomposition occupied most of the surface following laser-ablation. Following the silane treatment, its molecules formed Si-O-Al bonds with the rough surface. This markedly increased the number of C-C(H) hydrophobic groups on the surface. The superhydrophobic surface exhibited good self-cleaning, anti-icing and anti-corrosion properties. Synergistic interaction with the surface micro-nano structures and the large numbers of hydrophilic Al-OH/-OH groups which were generated by alumina-coating decomposition led to the surface superhydrophilicity. Following hydrophobic treatment, the silane molecules were grafted to the surface by Si-O-Al bonds and introduced massive C-C(H) bonds. The micro-nano structures on the superhydrophobic surface could trap air and prevent water droplets from touching the surface; thus improving the anti-icing properties by about 868%.

Table 7. Contact-angles and sliding-angles of water on AA6061 surfaces

Structure	Roughness (μm)	Contact-Angle (°)	Sliding-Angle (°)
original	0.09	54.1	>90
micro	2.78	148.4	13
nano	0.41	171.2	35
micro-nano	1.32	167.5	2.5

AA6061

A self-cleaning superhydrophobic AA6061 surface having a good corrosion resistance was created by using a combination of sand-peening and electrochemical oxidation, followed by coating with fluoro-alkylsilane[43]. The sand-peening pre-treatment introduced micro-scale pits. The electrochemical oxidation then produced nano-scale structures on the pits. The water contact-angle was 167.5°, with a sliding-angle of 2.5°. The effects of sand-peening parameters upon the surface wettability were studied in detail. These parameters included the impact-pressure, particle-size, nozzle-diameter and impact-time (table 6). The contact-angle of the original surface was 54.1° and the sliding-angle was more than 90° (table 7). Following sand-peening, the sliding-angle of the microstructure decreased to 13° but the contact-angle was 148.4°; indicating that the wettability of the microstructure had gone from hydrophilic to hydrophobic. Although water droplets on the microstructure could roll off, the surface had no superhydrophobic properties. The contact-angle of the nanostructure increased to 171.4° and the sliding-angle was 35°. Although this surface had a higher contact-angle, water droplets found it hard to roll off the surface. Potentiodynamic polarization curves were used to characterize the corrosion-rate of the alloy. A higher corrosion-potential indicated a lower corrosion-current density and a higher polarization resistance; reflecting a better corrosion-resistance. The corrosion-potential of the original surface was -0.679V (table 8), and the corrosion-current density was $0.0006249 A/cm^2$. The corrosion-potential of the microstructural surface was 16mV more positive than the original surface and the corrosion-current density decreased to $0.0001071 A/cm^2$. The corrosion-potential of the nanostructured surface was -0.634V, but it was inferior to that of the dual-structured surface. The associated current-density decreased to $0.000002087 A/cm^2$; 2 orders-of-magnitude lower than those of the original and microstructured surfaces. The corrosion-potential and current-density markedly improved when the sample surface became superhydrophobic. The corrosion-potential of the dual-structured surface was 87mV more positive than that of the original surface and 71mV higher than that of the hydrophobic microstructured surface. The associated corrosion-current density decreased by 3 orders-of-magnitude lower than that of the original surface. The overall data showed that the dual-structured surface had the lowest corrosion-rate and best corrosion-inhibition.

Table 8. Corrosion parameters for AA6061 surfaces

Surface	E_{corr} (V)	I_{corr} (A/cm^2)
original	-0.679	6.249×10^{-4}
micro	-0.663	1.071×10^{-4}
nano	-0.634	2.087×10^{-6}
micro-nano	-0.592	7.516×10^{-7}

An AA6061 surface with a contact-angle 154° and a sliding-angle of about 3.5° was prepared by electrochemical etching, surface-energy reduction and photolithography[44]. Many rectangular pits with sizes of a few microns were homogeneously distributed on the surface and micro-scale and nano-scale cavities were formed. A path with a radius of 50mm and a width of 400μm was created on the superhydrophobic surface by lithography. The contact-angles of water droplets on the path exhibited anisotropy, and 50μl water-droplets could closely slide along the path when the substrate was tilted by about 7°. The surface retained a good water-transport capability after 8 months. When the water droplet lay on the above path, the contact-angles perpendicular to, and parallel to, the path were anisotropic. Because of this, the droplet spread more easily along the line. The static contact-angle perpendicular to the path direction decreased to 98°, as compared with 149° along the path direction.

Due to its composition and microstructure, layered double hydroxides are ideal for creating self-healing superhydrophobic coatings. Such a Ni-Al layered double hydroxide coating was produced[45] on AA6061 by hydrothermal reaction and low-energy modification. The resultant coating, with nano-wall arrays, had a contact-angle of 162.1° and a rolling-angle of 1.9°. It exhibited excellent low-adhesion and self-cleaning properties, and could improve the corrosion-resistance. It could self-heal quickly under thermal stimulation after losing its superhydrophobicity. The wettability of the coating was governed by the microstructure and the surface energy. For a given modification method, the wettability was related to the microstructure. The microstructure and thus wettability were affected by the hydrothermal parameters.

Chemical etching and surface modification with very low surface energy 1H,1H,2H,2H-perfluoro-octyltriethoxysilane were used[46] to create superhydrophobic surfaces on AA6061 substrates. The contact-angle, rolling-angle and contact-angle hysteresis of surfaces etched with 8.0wt%HCl aqueous solutions were 162.5°, 1.9° and 1.1°, respectively. A contact-angle of above 150.0° could be retained for 90 days. The apparent surface free-energies of the superhydrophobic surfaces increased with decreasing surface temperature. The freezing-time of water droplets on the surfaces was delayed by 1568s, and the required temperature fell to as low as -11.9C.

A slippery biomimetic anti-icing coating was created by infusing silicone oil into superhydrophobic dual-scale micro-nano structured AA6061 surfaces[47]. The surfaces were prepared by combining chemical etching and anodization, followed by modification with polydimethylsiloxane. The oil-infused polydimethylsiloxane coating had an ice-adhesion strength of 22 kPa. The coating retained an ice-adhesion strength of 35kPa at -25C. The micro-nano structured surfaces exhibited excellent durability, with an ice-adhesion of only 108kPa following 20 icing/de-icing cycles. There was also a long-term icephobicity of 55kPa following 4 months of exposure to ambient environments. Thus the modified micro-nano structured surfaces possessed better anti-icing properties than did nano-structured surfaces under icing/de-icing and abrasion cycles. The micro-nano structured surfaces acted as a reservoir for holding excess oil, and reduced the loss of lubricant during icing/de-icing cycles. Without stable reservoirs, the lubricating layer was quickly depleted during repeated icing/de-icing cycles and the ice-adhesion strength increased. In order to explore the synergistic effect of polydimethylsiloxane and silicone oil on the coating, aluminium surfaces having various wettabilities were made by coating with polydimethylsilozane, 1H,1H,2H,2H-perfluoro-octyltrichlorosilane or silicone oil. The latter surface offered a much lower (22 kPa) ice-adhesion strength than did the others.

Table 9. Time to start-of-freezing of water droplets on various AA6063 surfaces

Surface	Temperature (C)
normal	-6
normal	-8
normal	-10
normal	-12
normal	-16
superhydrophobic	-6
superhydrophobic	-8
superhydrophobic	-10
superhydrophobic	-12
superhydrophobic	-16

Surface-nanostructuring and the chemical grafting of fluorocarbon molecules onto AA6061 substrates which had been laser-roughened and modified with nanostructured Al_2O_3 thin films were used to create a dual-roughness and porous surface states[48]. The surfaces were subjected to grafting with perfluoro-octyltriethoxysilane vapour, or were infused with low surface-tension liquid. A comparison of the wetting, water-condensation and anti-icing properties of the two systems showed that the grafted surfaces offered far better performances. The grafted surfaces were very superhydrophobic and required a higher water-vapour pressure in order to induce condensation. They also exhibited relatively long (4h) freezing-delay times for supercooled water droplets and markedly low ice-accretion in wind-tunnel tests.

The surface of AA6061 was rendered superhydrophobic by etching with hydrochloric acid, passivating with potassium permanganate and modifying with fluoro-alkylsilane[49]. With an etching-time of 360s and a passivation time of 3h, the resultant micro/nano-scale terrace-like hierarchical surface had a coral-like network structure. The surface had a water contact-angle of 155.7°, and an extremely weak adhesion to droplets. The corrosion inhibition in seawater was characterized by potentiodynamic polarization and electrochemical impedance spectroscopy. The superhydrophobic surface attained a corrosion inhibition-efficiency of 83.37%.

A simple method which combined droplet etching and chemical modification was used[50] to prepare a superhydrophobic AA6061 surface with a contact-angle of 156° and a sliding-angle of 5°. Unlike immersion, droplet etching could maintain the integrity of the aluminium and create a rough structure on its surface. Optimum superhydrophobicity was obtained by 1-step immersion in aqueous pentadecafluoro-octanoic acid solution at 80C. The surfaces exhibited thermostability – maintaining superhydrophobicity at 100 to 180C - plus anti-corrosion, self-cleaning and anti-fouling properties. The original and etched surfaces were hydrophilic, with contact-angles of 85.5° and 58.0°, respectively. An increase in roughness decreased the contact-angle and made the surface more hydrophilic. This explained why the surface became more hydrophilic due to droplet etching.

An easy method was found[51] for the preparation of a superhydrophobic AA6061 surface having a water-contact angle of 156° and a sliding-angle of 3°. The prepared surface had an hierarchical structure, and achieved a low surface-energy via a combination of sand-blasting and chemical modification. A study of the wettability after various immersion times indicated that 12h was the optimum immersion time for achieving the best superhydrophobicity. The resultant surface offered an excellent mechanical durability and could withstand abrasion with 1000-grit sandpaper under an applied pressure of 5kPa and a travel-distance of 800cm. The surface also exhibited good thermal stability when heated to 240C. It also offered good self-cleaning properties.

Table 10. Contact-angles and sliding-angles of water on AA6082 surfaces

Roughening	Silane	Contact-Angle (°)	Sliding-Angle (°)
-	no	69	-
-	yes	101	55
water	no	16	-
water	yes	174	>90
HNO_3/HCl	no	23	-
HNO_3/HCl	yes	160	26
HF/HCl	no	12	-
HF/HCl	yes	179	0

AA6063

A superhydrophobic AA6063 surface having improved enhanced anti-icing and self-cleaning performances was produced[52] by immersion in N,N-dimethylformamide – stearic-acid solution. The anti-icing behaviour was explored by measuring the frosting/icing time-delay, the freezing-temperature reduction and the frost/ice adhesion. Simulated polluting particles could be completely removed by rolling water-droplets. Such droplets, with a volume of 10µl, on the as-prepared surface had a quasi-spherical shape and the contact-angle could be 155.5° while the rolling-angle was less than 4°. Droplets could roll downhill quickly when the surface was tilted by 3 to 5°. The as-prepared superhydrophobic alloy had a rough surface with a micro-nano scale structure, and hydrophobic alkyl chains grafted onto the surface. Any water-droplet on the surface was thus in contact with the low surface-energy film and with air. The contact area between the water droplet and air accounted for about 92% of the total area. The superhydrophobic surface greatly hindered the formation and growth of frost crystals, and melted crystals soon became rolling water-droplets. The surface could delay the freezing-time by 5 to 9min or reduce the freezing-temperature by 2 to 4C (table 9). The rolling water-droplets could entirely carry away pollution particles on the superhydrophobic surface. The as-prepared surfaces could trap a large amount of air. The trapped air could repel water and vapour at low temperatures due to a cushioning and heat-insulation effect. Nucleation and crystal growth of ice and frost on the superhydrophobic alloy surfaces were then lower than those on normal surfaces; whence the excellent anti-icing properties.

AA6082

A 2-stage method was used[53] to produce superhydrophobic AA6082 surfaces was proposed in which the first step involved the creation of a rough nano-micro structure and the second step involved the reduction of surface energy by using octadecyltrimethoxysilane. The roughening was achieved by short-term pre-treatment with boiling water, HNO_3/HCl etchant or HF/HCl etchant. The surface energy was reduced by dip-coating with a dilute solution of octadecyltrimethoxysilane so as to form self-assembled silane monolayers on the alloy surface. The latter always had an hierarchical nano-micro roughness. The wettability depended upon the roughening pre-treatment: a very high water contact-angle and a low sliding-angle were found for a HF/HCl-etched silanized surface (table 10). There was a marked increase in the corrosion-resistance to 3.5wt%NaCl, and this increased with decreasing sliding-angle. Electrochemical parameters such as the corrosion current density, I_{corr}, and the corrosion potential, E_{corr}, were measured (table 11). A lower corrosion current density and a higher corrosion potential indicate a superior corrosion resistance.

Table 11. Open-circuit potential values of AA6082 surfaces in 3.5wt%NaCl solution

Roughening	OCP (V)	E_{corr} (V_{SCE})	I_{corr} ($\mu A/cm^2$)
-	-0.931	-0.906	21.17
-	-0.786	-	-
water	-0.769	-	-
water	-0.698	-0.735	0.04
HNO_3/HCl	-0.645	-	-
HNO_3/HCl	-0.537	0.449	0.03
HF/HCl	-0.620	-	-
HF/HCl	-0.472	-0.419	0.03

AA7075

Biomimetic superhydrophobic surfaces with a micro-nano hierarchical structure were produced[54] on AA7075 by combining laser-ablation and anodic oxidization, leading to a contact-angle of 164° and a sliding-angle of 2°. Arrays of lotus-leaf like structures existed on the surface, with the distance between the centres of motifs being 100μm. A mechanical model was used to analyse the dynamics of super-cold water droplets, having various pH-values, impacting the cold superhydrophobic surface. Dynamic analysis indicated that the surfaces exhibited differing surface adhesions, depending upon the pH-value of the droplet.

AMG

The change in the wettability of alternative superhydrophobic coatings AMG (Al-2.9wt%Mg) was studied[55] following extended contact with 3M potassium halide solution. The corrosion currents and rates of hydrolysis of hydrophobic molecules respected the Hofmeister order: $I^- < Br^- < Cl^-$. The uppermost layers of both coatings had a very similar morphology. The perfection and stability of the layer of hydrophobic molecules on the textured surface was considered to be a determining factor in corrosion protection. The coatings differed in hydrolytic activity, phase composition and high-temperature oxide-content.

AMS4037

Superhydrophobic surfaces with an hierarchical structure were created[56] on AMS4037 aluminium alloy plate by etching with dilute hydrochloric acid and hydrogen peroxide. When compared with untreated alloy, the superhydrophobic surface exhibited an excellent water-resistance, with a water contact-angle of up to 163.6° plus a minimal rolling-angle and contact-angle hysteresis. Anti-icing tests showed that the weight of ice on the superhydrophobic surface was much lower than that on other aluminium surfaces for a given test period. The weight of ice on the treated material was 1.7751g, as compared with 1.9952g on the untreated surface when the test period was 2h.

Cobalt

Superhydrophobic cobalt surfaces with a micro-nano fibre structure were prepared by using a 1-step electrodeposition process[57]. Coatings which were electrodeposited at 20V for 300s were superhydrophobic, with a contact-angle of 160° and a sliding-angle of 6.2°. The superhydrophobicity arose from the form of cobalt myristate on the surface, together with the micro-nano fibre structure. The use of an excessive electrodeposition-time, or voltage, changed the micro-nano fibre structure and decreased its superhydrophobicity.

The surface could be changed from superhydrophobic to hydrophilic by heating, and then dipping in an ethanoic solution of 0.1M myristic acid. The prepared coating offered a better corrosion-resistance than did the bare substrate. A corrosive medium of strong alkalinity or acidity decreased the corrosion-resistance, and the corrosion-resistance was consistent with the hydrophobicity.

An amphiphobic cobalt coating with a contact-angle of 121° was created[58] by optimizing the electrodeposition conditions to a current-density of 92mA/cm^2 and a pH of 3.25. The hydrophobicity was attributed to an hierarchical morphology which was produced by deposition, together with the gradual removal of molecular water and the adsorption of environmental hydrocarbons over a period of 120h. The coating was then modified by using a fluorinated solution. The surface then became superhydrophobic, with a water contact-angle of 161° and a contact-angle hysteresis of about 8°. The roughness of the modified coating remained constant, and the increased liquid-repellency was ascribed to the low surface energy. The modified coating increased corrosion-inhibition by reducing the corrosion-current density by 82%. The effect of current density upon hydrophobicity guided the search for the optimum deposition conditions with regard to contact-angle.

One-step electrodeposition was used[59] to synthesize superhydrophobic cobalt coatings having a water contact-angle of 155.6° and a sliding-angle of 4.8°. The coating was highly corrosion-resistant, with a charge-transfer resistance of 756.3kΩcm^2, and was consistent with its stability during 16 days of immersion in 3.5wt%NaCl solutions with a pH-value of 1 or 14. It remained hydrophobic, with a contact-angle of greater than 90°. Water droplets on the coating, at -15C, required a 23 min longer freezing time, as compared with that on the substrate. In addition, just 3 rolling water drops were required to remove alumina powder from 1.13cm^2 of the coating.

Copper

Superhydrophobic surfaces on 2N5-purity copper were obtained[60] by combining etching in 10wt% ammonia solution and calcination at 340C. The surface was further modified using an ethanol solution which contained stearic acid. The resultant surfaces had contact-angles as high as 157.6° and exhibited a persistent corrosion-resistance in 3.5wt%NaCl aqueous solution. The untreated copper surface was very smooth, with a contact-angle of 76.5°. Following etching for 20h, there were irregular cell-like projections with a height of several micrometres on the surface. That surface had a contact-angle of 21°.

A 30V direct-current voltage was applied between two copper plates, 1.5cm apart, immersed in dilute ethanolic stearic acid solution[61]. The surface of the copper anode became superhydrophobic due to a reaction between the copper and the stearic acid, such that it became covered with flower-like low surface-energy copper stearate. This imparted a water contact-angle of 153°, and roll-off properties.

Wet chemical reaction was used to create a superhydrophobic surface on a polished copper substrate at room temperature[62]. The surface had a water contact-angle of about 154° and a sliding-angle of about 4°. These were attributed to the roughening caused by the chemical reaction and to the low surface free-energy which was produced by treatment with vinyl-terminated polydimethysiloxane. Copper oxalate features with an average diameter of about 0.5μm, and circular sub-microscopic structures with a diameter of about 100nm, made up an hierarchical structure comprising micro- and nano-scale elements. The maximum peak-to-valley and root-mean-square values were about 120nm and about 15.4nm in size, respectively.

Etching and hydrothermal treatment were used[63] to produce a superhydrophobic surface on copper with a contact-angle was 157.7°. The bare copper surface had a contact-angle of 76.5°. A few microns of pebble-like structure appeared on the etched copper surface, and the contact-angle was only 21°, thus indicating that the copper surface was uniformly dissolved in the $NH_3 \cdot H_2O$ solution. The coating was the stearic acid salt of copper, which was formed by the reaction of stearic acid with cuprous oxide. The surface exhibited a corrosion-inhibition efficiency of 99.81% in 3.5wt%NaCl aqueous solution, and good stability in simulated seawater and humid air.

The ability of superhydrophobic surfaces to remain dry is useful for frost suppression on metal surfaces but, when subjected to pressure, the protective layer is easily damaged and this leads to a loss of superhydrophobicity. Surfaces of 3N-purity copper were modified[64] using nanosecond laser and 1H,1H,2H,2H-perfluoro-octyltri-ethoxysilane or re-filled nano-silica. The inverted-pyramid microstructures were arranged in a continuous regular pattern, with the outer layer constituting a compressive wear-resistant surface and the inner layer being filled with hydrophobically-treated nano-silica. The contact-angle of the surface was 160.3° and the rolling-angle was 1°. The mechanical resistance of the superhydrophobic surface was tested using knifes, stainless-steel wire and tape. The freezing and anti-freezing behaviours of droplets on the superhydrophobic copper, and on ordinary copper, were compared, showing that the superhydrophobic surface retained its superhydrophobicity following repeated durability tests. Unlike superhydrophobic coatings which were applied directly to the substrate, the combined surface did not lose its hydrophobicity in the case of partial wear. Because the outer layer of the combined surface had a micron-sized framework it protected the fragile nanostructure within. When the frost-thickness was 0.9mm, at a cooling temperature of -7C, the horizontal superhydrophobic surface exhibited excellent frost-suppression, as compared to that of the ordinary surface, and frost-growth could be delayed by 1.75 times. The hydrophobicity of the superhydrophobic surface remained essentially the same following 50 freeze-thaw cycles with a cooling-time of 0.5h.

Copper-based superhydrophobic materials were prepared by means of oxidation, lauryl mercaptan-modification and compression moulding[65]. The surfaces had a cauliflower-like structure, with long lauryl mercaptan chains self-assembled onto them. The contact-angle was 155.2° and the sliding-angle was less than 5°. The surfaces could be exposed to air for 10 months, soaked in water for 2 weeks and suffer pH-values ranging from 6 to 14 without any notable change. The superhydrophobicity could also be restored if the surfaces were subjected to abrasion or scratching. The material also exhibited excellent self-cleaning capabilities because contaminant particles could be easily swept away by rolling water-droplets.

A 1-step electrodeposition method was used to create a micro-nano superhydrophobic structure on the surface of copper by using a choline chloride based ionic liquid as the electrolyte[66]. As compared with bare copper, with a contact-angle of 62.2°, the contact-angle of the electrodeposited superhydrophobic coating attained 157.8°. The corrosion-resistance behaviour of the coating in 3.5wt%NaCl solution was extremely good.

Superhydrophobic 4N6-purity copper surfaces were prepared by means of an oxidation method involving $NaClO_2$, NaOH and $Na_3PO_4 \cdot 12H_2O$, followed by modification with 1-octadecanethiol[67]. The optimum conditions were an oxidation time of 0.5h, a modifier concentration of 2.5mM and a modification time of 5h, leading to a contact-angle of 161.1° and a sliding-angle of 2.2°. The surfaces were immersed in de-ionized water, and exposed to air for several days. The contact-angle was reduced to 119° by immersion in water for 1 week, but was not affected by air-exposure after 45 days.

Superhydrophobic copper was prepared by using a 2-step process of electrodeposition followed by behenic acid coating, or a 1-step coating process without the electrodeposition[68]. The contact-angle was 151° for the 1-step process and 153° for the 2-step process. The robustness of the coated copper was evaluated by abrasion with sand-paper, immersion in acidic or alkaline solutions and heating to high temperatures. The wetting behaviour of the coating was impaired by all of those tests, but the corrosion behaviour did not follow a similar trend. The 2-step process was more efficient than the 1-step process.

A template and etching method was used to create a regular hierarchical multi-scale structure on copper foil by using the surface of bamboo leaf as the template[69]. This structure increased the water contact-angle of the foil surface from 64° to 131.1°. The hierarchical structure was then further modified using stearic acid, leading to a contact-angle of 160.0° and a sliding-angle of 3°.

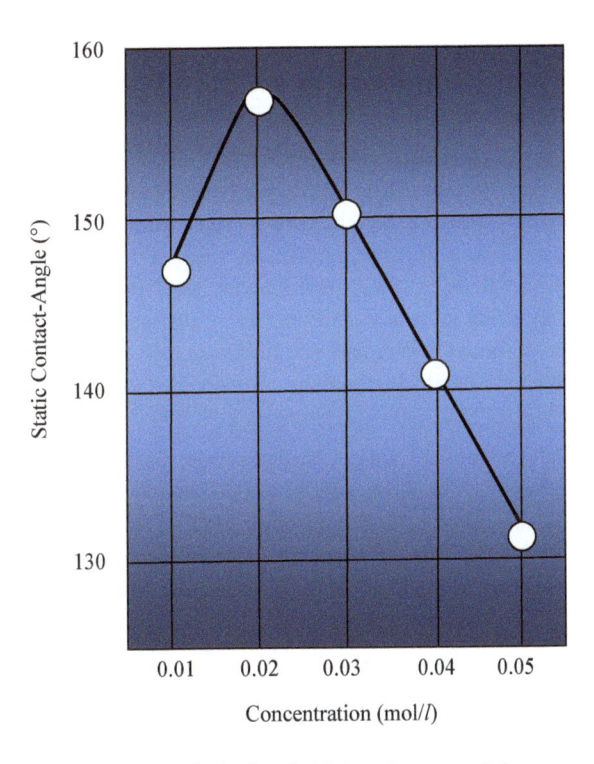

Figure 1. Influence upon the hydrophobicity of copper of the concentration of octadecanoic acid in ethanol solution

Chemical etching was used to produce superhydrophobic copper surfaces by initial immersion in ferric chloride solution[70]. The etched surfaces had a maximum contact-angle of 140°. They had high a sliding-angle, and water droplets were retained even on inverted surfaces. Following stearic acid modification of the etched surfaces, the contact-angle increased to above 150° and the sliding-angle decreased to less than 10°.

Coatings with copper deposits were prepared by using jet-electrodeposition[71]. The coatings had a micro-nano structure. Following modification with stearic acid, a superhydrophobic surface was obtained. The static contact-angle and the sliding-angle of the surface were 151.6° and 5.7°, respectively. The coating had a higher corrosion-potential and lower corrosion-current than those of bare copper.

A superhydrophobic copper surface was created by using electrochemical deposition and lauric acid functionalization[72]. The surface had contact-angles as high as 158°, and exhibited very good self-cleaning, anti-icing and drop-wise condensation capabilities. The thermal stability range was -15 to 150C, and the coating resisted abrasion, and acidic or basic media.

Periodic surface ripples were produced on 3N-purity copper by using picosecond laser (1064nm wavelength, 203.6kHz repetition-rate, 10ps pulse-width) nanostructuring. Following modification with triethoxyoctylsilane, various types of ripple exhibited differing levels of wettability[73]. Fine ripples, with few re-deposited nano-particles, exhibited a high attraction to water. An increased amount of nano-scale structure decreased the adhesive force to water and also increased the contact-angle. One specific type of ripple exhibited superhydrophobicity, with a contact-angle of 153.9° and a sliding-angle of 11°.

Nanosecond laser-processing and sol-gel methods were used to produce micro-nano inverted-pyramid structures, modified with SiO_2-polydimethylsilane, on copper[74]. The superhydrophobic surface had a water contact-angle of 159.5° and a sliding-angle of 0.5°. The surface could efficiently remove contaminants within 200s. In anti-icing tests, the delayed icing-time and ice-adhesion strength of the surface were 3.33 times and 1.99% those of ordinary copper surfaces, respectively. In electrochemical tests, the corrosion current density decreased by an order of magnitude, and the corrosion-inhibition efficiency attained 90.57%. The middle SiO_2-polydimethylsilane coating conserved superhydrophobicity when the top-most nanostructure was worn. As wear continued, the inverted-pyramid micro-nano structural array protected the internal nano-hydrophobic material. The adhesion of water to the surface was 3.27μN.

Superhydrophobic surfaces were produced on copper plate by treatment with $AgNO_3$ and dodecyl mercaptan[75]. The as-prepared surfaces had a hierarchical rough structure which comprised nano-sheets and nano-particles. Long alkyl chains were assembled on the rough surface. This led to a water contact-angle of 156.8° and a rolling-angle of about 3°. The surface exhibited long-term durability and excellent stability. Its impressive performance was attributed to the so-called cushion-effect and to capillary phenomena. water and aggressive liquids could thus be prevented from contacting the copper surface, and contaminants could be easily washed away by rolling water droplets. Icing was also delayed by the superhydrophobic surface.

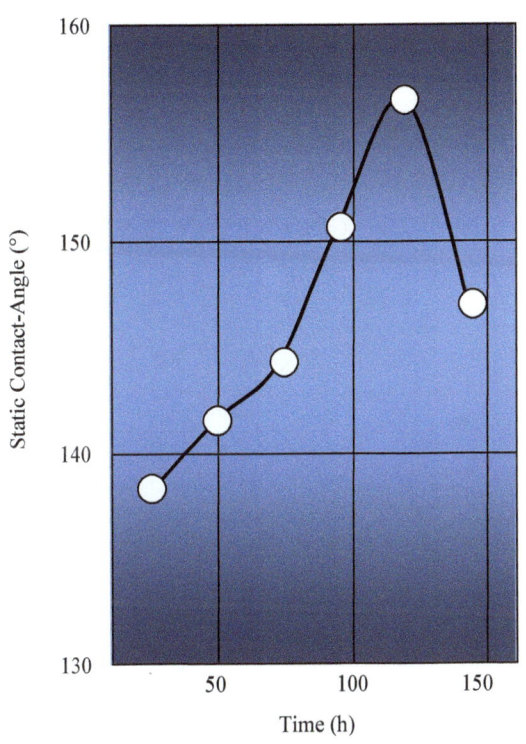

Figure 2. Influence upon the hydrophobicity of copper of the immersion-time in octadecanoic acid solution

The wetting-transition from the Cassie mode to the Wenzel mode degrades the performance of superhydrophobic surfaces. The wetting-transition occurs when the surface tension can no longer resist the gravitational force, and the liquid penetrates the spaces between asperities; leading to collapse. The wetting stability of 3N-purity copper-based superhydrophobic surfaces was therefore investigated[76]. The samples had nano-asperities with a diameter of 70nm, but with 2 different packing-densities. The static (sessile-droplet) and dynamic (drop-wise condensation) wetting stabilities were compared. Sessile droplets on surfaces with densely-packed nano-asperities having a pitch of 120nm remained in the stable Cassie mode. The wetting-transition from Cassie mode to Wenzel mode occurred spontaneously on coarsely-packed nano-asperities having a pitch of 300nm. The contact-angle on the surfaces of coarsely-packed nano-asperities decreased from over 150° to about 110°, and the sliding-angle increased from less than 5° to over 60° within 200s. There were essentially no changes on the other surface. With regard to drop-wise condensation, condensed droplets on surfaces with densely-packed nano-asperities remained in stable Cassie mode, whereas the droplets on surfaces with coarsely-packed nano-asperities were in Wenzel mode.

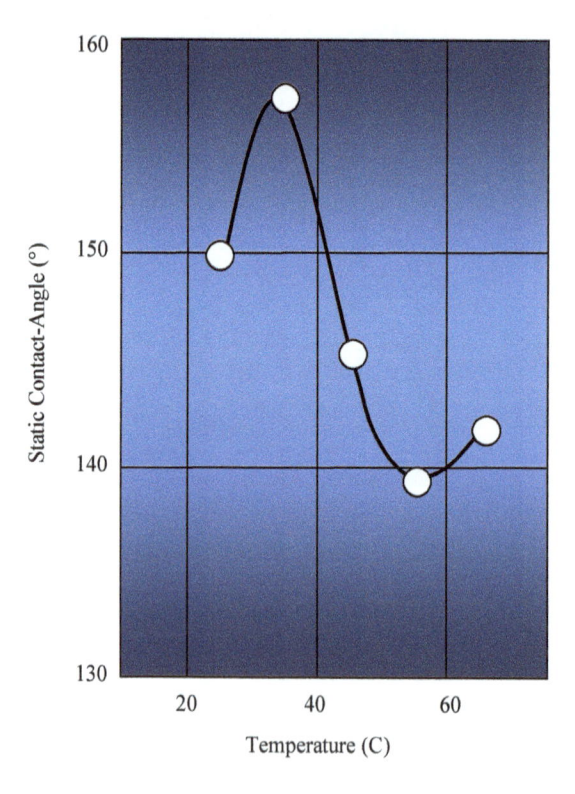

Figure 3. Influence upon the hydrophobicity of copper of the immersion temperature in octadecanoic acid solution

Hierarchical micro-nano scale binary rough structures were created on copper by 30V direct-current electrochemical machining in a neutral 0.2mol/*l* NaCl electrolyte[77]. It required only 3s to produce the necessary roughness. The rough structures comprised micrometre-scale potato-like features and nanometre-scale cube-like features. Following modification with fluoro-alkylsilane, the surfaces were superhydrophobic, with a water contact-angle of 164.3° and a tilting-angle of less than 9°.

Copper sheet was etched in 25wt% ammonia solution, under ultrasound, so as to produce a uniform roughness, and then immersed for 120h in a 0.02mol/l solution of octadecanoic acid in 35C ethanol before annealing (120C, 1h) to give a superhydrophobic coating (figures 1 to 3)[78]. The ultrasonic treatment shortened the etching-time and improved etching uniformity. The static contact-angle was up to 157°, and the contact-angle hysteresis was 4.2°. The superhydrophobic improved the anti-corrosion and self-cleaning properties. Due to the coating, the corrosion-current was decreased to 6.8911 x 10^{-6}A/cm² from 5.5577 x 10^{-5}A/cm² and the corrosion-potential was increased to -0.1966V from -0.2878V, indicating an improved corrosion-resistance. The proposed mechanism was that copper atoms reacted with octadecanoic-acid molecules and a long carbon-chain structure formed on the surface, imparting superhydrophobicity and keeping corrosive ions away.

Table 12. Surface energy of milled copper

Cutter Tip Distance (mm)	Energy (mN/m)
-	47.296
25	0.324
30	0.259
35	0.225

Milling, deposition of $AgNO_3$ solution and modification with stearic acid were used[79] to create superhydrophobic copper surfaces (table 12). The surface morphology was dendritic and rectangular surface promontories, produced by the milling, were distributed over the substrate. The water contact-angle could be as high as 158.4°. The best anti-corrosion behaviour was exhibited when the milling was conducted using a cutter tip distance of 0.30mm (table 13). The as-prepared superhydrophobic surfaces offered good self-cleaning. When scratched with a knife and abraded, the substrate retained a good superhydrophobicity. The coating was mechanically stable and had a good corrosion resistance. The I_{corr} and E_{corr} data, with their low positive values, indicated that treatment improved the corrosion-resistance of the copper substrate. The wettability of the surface after wear-tests showed that, after being pulled 5 times, the contact-angle was 152.7°;

lower than the value of 157.6° before the friction test. After pulling 10, 15 or 20 times, the contact-angle remained above 148°. After pulling 20 times, the contact-angle was 148.2°. The surface followed the Cassie model, with a high contact-angle and low viscosity.

A superhydrophobic surface was produced on copper[80], as evidenced by a water contact-angle 147° and a roll-off angle of 5°. The surface was created by annealing copper foil in air and coating it with silica nanoparticles that were dispersed in a silane solution. There was a uniform outer distribution of spherical micron-sized CuO particles over the whole surface and an inner layer comprising a mixture of CuO and Cu_2O. Siloxane-bonding to the substrate was detected. The silane-coated surfaces exhibited augmented valleys and peaks having a higher root-mean-square and average roughness, due to the silica nanoparticles. Electrochemical studies in aqueous chloride environments revealed a corrosion-resistance, as reflected by a shift in open-circuit potentials in the noble direction, by an increase in the charge-transfer resistance and by a lower anodic current, as compared with those of as-received copper foil. The critical surface energy of the superhydrophobic surface was calculated to be 17.72mN/m. The coating remained stable after 8 months of immersion in de-mineralised water under ambient conditions.

Table 13. Electrochemical corrosion data for milled copper surfaces

Cutter Tip Distance (mm)	E_{corr} (V_{SCE})	I_{corr} (A/cm^2)	Corrosion Rate (mm/a)
-	-0.2208	2.92×10^{-5}	0.3440
25	-0.1354	1.80×10^{-5}	0.2100
30	-0.1492	9.87×10^{-6}	0.1150
35	-0.1236	1.88×10^{-5}	0.2187

Sandblasting and hot water treatment were used to create micro-roughness and nano-roughness, respectively, on copper[81]. Bare sheets were sandblasted using Al_2O_3 particles, while the hot water treatment involved simple immersion in de-ionised water at water at 75C for 24h. The processing resulted in copper oxide nanostructures, formed by the hot water treatment, coated onto the microstructured surface which was produced by sandblasting. The nanostructures possessed CuO stoichiometry and had the form of leaves, with a thickness of 15nm and a width of 250nm. The nano-leaves, when located at the side-walls of micro-hills, seemed to provide a overhang topography which was critical to imparting superhydrophobicity. The hierarchically rough samples were then coated with 1H,1H,2H,2H-perfluorodecyltrichlorosilane, which reduced the surface energy. The resultant water contact-angles were as high as 164°. Fluorinated control, nano-rough and micro-rough surfaces had a range of other angles (table 14). Some non-fluorinated micro-nano rough surfaces exhibited strongly hydrophilic behaviour, with a contact-angle of 50°. Durability tests showed that the superhydrophobic surfaces could offer good wetting stability, due to their self-cleaning properties. The morphology and crystal structure of the copper oxide nanostructures were stable under moderate annealing conditions.

The anti-corrosion properties of superhydrophobic copper surfaces were investigated following wet chemical etching and immersion treatments[82]. The corrosion-resistance was characterized after immersing surfaces in 3.5wt%NaCl solution (table 15). Micro-scratches and grooves were visible on the pristine surfaces. At the nano-scale, the surfaces appeared leaf-like. These sharp dense structures contributed to the high roughness which was associated with the substrates. The surface contained copper and oxygen, suggesting that the leaf-like nanostructures on the surfaces were CuO. The average surface roughness of the surfaces was 428nm. The average water contact-angle was 83° for the pristine copper surface, 110° for the copper surface when coated with trichlorosilane and 169° for the superhydrophobic surface.

Table 14. Contact-angles of variously processed copper surfaces

Morphology	Fluorinated	Contact-Angle (°)	Nature
micro-nano rough	no	20.5	strongly hydrophilic
micro-rough	no	60.0	strongly hydrophilic
nano-rough	no	73.0	strongly hydrophilic
flat	no	78.0	hydrophilic
flat	yes	132.0	hydrophilic
nano-rough	yes	137.0	hydrophilic
micro-rough	yes	149.3	hydrophilic
micro-nano rough	yes	164.0	superhydrophilic

Superhydrophobic copper surfaces with a contact-angle of 156.2° and a sliding-angle of 4° were prepared by means of hydrothermal treatment and silane modification[83]. The superhydrophobic Cu_2S-coated surface comprised a large number of Cu_2S crystals grafted to long hydrophobic alkyl chains. The material thus had a rough hierarchical surface with micro- and nano-scale features. The surface exhibited good mechanical durability and corrosion resistance, plus self-cleaning. Heat treatment at 200C had a marked effect upon the surface microstructure and wettability. The mechanical durability and stability of the superhydrophobic copper were evaluated by means of abrasion tests with 1500-mesh silicon carbide paper as the abrasive surface. The sample surface was dragged over various distances under a pressure of 1.25kPa. The water contact-angle and sliding angle were measured after dragging (figure 4). The corrosion-resistance was evaluated using electrochemical corrosion testing (table 16) and long-term immersion in 3.5wt%NaCl solution at 25C for up to 65 days.

Table 15. Potential values of copper surfaces in 3.5wt%NaCl solution

Condition	E_{corr} (V)	I_{corr} (A/cm^2)	Corrosion Rate (mm/y)
pristine	-0.220	2.12×10^{-6}	4.89×10^{-2}
trichlorosilane-coated	-0.17	6.6×10^{-7}	1.53×10^{-2}
superhydrophobic	0.106	4.62×10^{-9}	1.07×10^{-4}

Superhydrophobic copper surfaces were prepared by oxidation, heat-treatment and alkyl-chain grafting[84]. The surfaces had a structure which comprised CuO nano-sheets and needle-like fibres. The micro-nano scale hierarchical surface and grafted long alkyl chains imbued the surface with water-repellence, and the water contact-angle and sliding-angle could attain 157.3° and 5°, respectively, after modification with stearic acid. The contact-angle gradually increased with increasing immersion time up to 24h. It then decreased with increasing immersion time. This was attributed to Cassie-Baxter behaviour. When treated in stearic acid ethanol solution for less than 24h, the alkyl chains grafted onto the surface. During continued immersion, more stearic acid molecules were grafted. More and more air was therefore trapped in the solid/liquid contact area under a water droplet. When the immersion time was too long, the number of grafted stearic acid molecules saturated. Some molecules were then deposited on the surface in the form of physical adsorption, causing the contact-angle to decrease. The corrosion behaviour in 3.5wt%NaCl aqueous solution was studied. The E_{corr} value of the plain copper was -

3.60V, while the E_{corr} value of the superhydrophobic copper increased to -1.50V. The I_{corr} values of the plain and superhydrophobic copper were 5.248 x 10^{-5} and 5.623 x 10^{-6} A/cm^2, respectively. As usual, a lower corrosion current density or a higher corrosion potential implied a lower corrosion-rate and a better corrosion resistance. The increase in E_{corr} and decrease in I_{corr} for superhydrophobic copper was attributed to the so-called cushion effect and to capillarity. The hierarchical surface structure could easily trap a large amount of air at the interface, which then stopped water and chlorine ions from contacting the surface and; thus preventing corrosion.

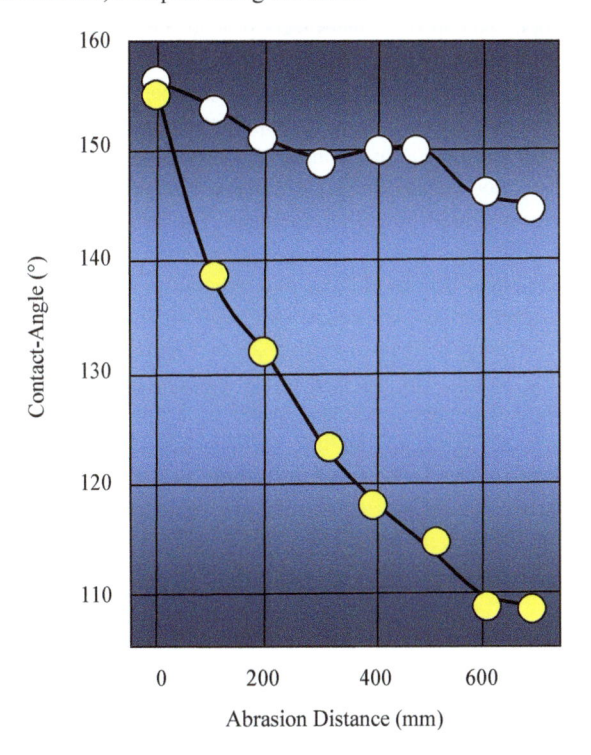

Figure 4. Effect of abrasion distance upon the contact-angle of copper. Yellow: without 200C heat treatment, white: with 200C heat treatment

Table 16. Corrosion potential and current density of copper in 3.5wt%NaCl solution

Condition	E_{corr} (V)	I_{corr} (A/cm^2)
plain	-0.72	2.34×10^{-5}
superhydrophobic	0.14	2.29×10^{-8}

Stable superhydrophobic copper surfaces were produced by using 1062nm laser-ablation to produce line and grid patterns[85]. Ablation increased the surface roughness, increased the number of (111) planes and decreased the number of (200) planes. The rapid evolution of the patterned surface from hydrophilic to superhydrophobic caused a rapid change from hydrophilic CuO to hydrophobic Cu_2O and organic adsorption. The wetting properties could be modified by changing the step size, and the stability of the water contact-angle was monitored for up to 120 days. The raw surface had a hydrophobic contact-angle of 119°, but the angle changed to hydrophilic on all surfaces following laser-ablation. The angle decreased from hydrophobic to hydrophilic due to the CuO structure on the copper surface. The contact-angles of the line and grid patterns generally increased upon increasing the step-size from 20 to 150μm; apart from the step-size of 150μm for the line pattern. The angle increased because of the raised non-ablated and hydrophobic surface. The angle decreased to 43° on a line pattern with a 150μm step-size, and this was attributed to the line patterns having an irregular ablation-spot distribution. The contact-angle increased with time such that, after 5 days, the surface changed from hydrophilic to hydrophobic for line patterns with 100μm steps or grid patterns with 20 or 100μm steps (table 17).

Table 17. Contact-angles of laser-patterned copper surfaces

Pattern	Step-Size (μm)	Condition	Contact-Angle(°)
line	20	ablated	16
line	20	aged, 5days	105
line	100	ablated	48
line	100	aged, 5days	155
line	150	ablated	43

Pattern	Step-Size (μm)	Condition	Contact-Angle(°)
line	150	aged, 5days	117
grid	20	ablated	20
grid	20	aged, 5days	152
grid	100	ablated	40
grid	100	aged, 5days	152
grid	150	ablated	57
grid	150	aged, 5days	111

When the step size of the line pattern and the grid pattern was 150μm, the contact-angle decreased to 117° and 111°, respectively. The decrease in angle with increasing step size was attributed to easier droplet contact with flat areas. A line pattern with a step size of 100μm had the highest contact-angle because that pattern had the highest height/diameter ratio, and the latter determined the effectiveness of air-pocket formation under the water surface. An increase in depth led to an increased angle because the droplet could not touch the bottom of the valley. When the depth was not too deep, the droplet touched the bottom and that promoted Wenzel behaviour at the surface. Following same-day annealing (110C, 2.5h), the contact-angle increased from less than 70° to greater than 150° (table 18); apart from a line pattern with a 20μm step-size. The wettability in the latter case changed from hydrophilic to hydrophobic surface because air-pockets could occur only in the main groove of the ablation area.

Table 18. Contact-angles of laser-patterned copper surfaces

Pattern	Step-Size (μm)	Condition	Contact-Angle(°)
-	-	ablated	119
-	-	heat-treated	111
line	20	ablated	16
line	20	heat-treated	139
line	100	ablated	48
line	100	heat-treated	157

line	150	ablated	43
line	150	heat-treated	155
grid	20	ablated	20
grid	20	heat-treated	157
grid	100	ablated	40
grid	100	heat-heated	160
grid	150	ablated	57
grid	150	heat-treated	163

When the step size became small, deformed copper covered the protuberant part and welded the island so that the surface then had a semi-regular pattern. The contact-angle on grid patterns was more stable because those patterns had a small portion of flat surface, surrounded by considerable trapped air which prevented liquid from wetting the surface. Following heat treatment, the contact-angle on textured surfaces had a greater. The laser-ablated textures exhibited a lower adhesion-force and sliding-angle than those of the raw surface. A decreasing adhesion-force favoured a low sliding-angle (table 19).

Superhydrophobic surfaces with dense nanostructures were created on copper substrates by means of template-assisted electrochemical deposition[86]. During deposition, the bubbles which were generated persisted in some regions and prevented the creation of the nanostructure, thus producing an heterogeneous surface. In order to assure an homogeneous surface, the electrolyte composition and the voltage had to be optimized. Condensation heat-transfer on superhydrophobic surface was studied, and the heat-transfer performance was found to deteriorate during experiments. Compared with hydrophilic and smooth surfaces, the heat-transfer coefficient on fresh superhydrophobic surfaces was improved by up to 154.3%, but the heat-transfer coefficient of superhydrophobic surfaces, after repeated tests, offered only a 67.2% improvement at best. There was no obvious change in the nanostructure following repeated experiments, but the polymer which was attached to the nanostructure and which imparted hydrophobicity was destroyed.

Table 19. Adhesion force and sliding-angle of laser-patterned copper surfaces

Pattern	Size (μm)	Condition	Adhesion Force (μN)	Sliding-Angle(°)
-	-	-	223.1	70
-	-	heat-treated	224.4	70
line	100	aged, 5 days	64.4	15
line	100	heat-treated	61	15
grid	100	aged, 5 days	59	4.3
grid	100	heat-treated	50	4.3

Iron

Wire-like ZnO particles which were micro-scale in length and nano-scale in width were deposited onto an iron surface by means of ultrasound[87]. Stearic acid in ethanol solution was then used for surface-energy reduction. The water contact-angle was 158.47 and the lowest sliding-angle was 7.66°. Potentiodynamic data indicated that, when ultrasound was used for the particle deposition, the corrosion current was 1000 times smaller than that for the bare iron. The usual method for particle deposition had no marked effect upon the corrosion resistance. When ultrasound was used for particle deposition, the stability of the resultant superhydrophobic iron during immersion in 3.5wt%NaCl solution was greater than that of superhydrophobic iron which was prepared by using conventional techniques.

Hierarchical structures were prepared on iron surfaces by chemical etching with hydrochloric acid, or by galvanic treatment in silver nitrate solution, and then modified using stearic acid[88]. The latter chemically bonded to the iron surface. As the HCl concentration was increased from 4 to 8mol/l, the surface became rougher, and the water contact-angle changed from 127° to 152°. The nitrate concentration had little effect upon the wetting behavior, but a high concentration caused silver particle aggregates to change from flower-like to dendritic, due to the preferential growth direction of the silver. When compared with etching, the galvanic replacement method was more favourable to producing the roughness required for superhydrophobicity. The iron surface offered excellent anti-icing properties, as compared with those of untreated iron. The icing-time of water droplets on the superhydrophobic surface was delayed to 500s, as compared with the 295s for untreated iron. The former surface retained its superhydrophobicity after 10 icing and de-icing cycles.

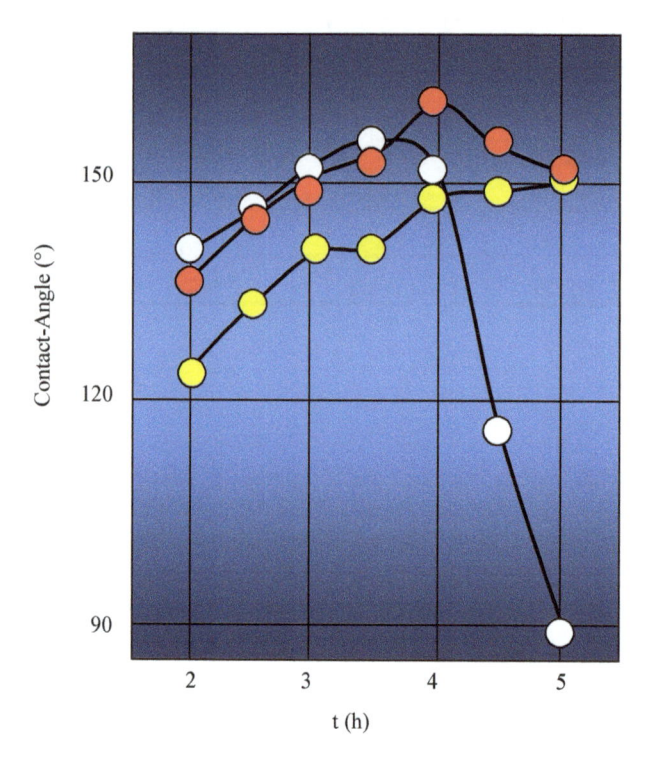

Figure 5. Contact-angles of iron surfaces as a function of annealing time. Yellow: 400C, red: 450C, white: 500C

Superhydrophobic surfaces were prepared on iron by means of heat-treatment and surface-energy modification[89]. Micron-level structures were first constructed on the surface by scribing, followed by thermal oxidation to generate nanostructures upon the micron structures. The contact-angle was increased from 9.6° to 89.8° or 128.0° by scribing in various ways. The oxidation treatment changed the contact-angle to 21.7°, 122.2° and 138.1° for ordinary iron, iron scribed horizontally and vertically scribed and surfaces scribed multi-directionally, respectively. Following modification using silanes, the contact-angles were 107°, 131.1° and 157.3°. The greatest water contact-angle was 162.3°, together with a sliding-angle of 2.4° after annealing (figure 5) under the optimum conditions (450C, 4h). When the annealing temperature was 500C, the surface was unstable and the oxide layer peeled off. These surfaces rendered the iron highly water-repellent and imparted anti-condensation behaviour together with good mechanical stability and corrosion resistance. The low surface energy of the coating also led to a poor adhesion to dust and bacteria. The above process was applicable to surfaces containing over 90% of iron. When the surface was too hard and wear-resistant it was difficult to carry out the required surface roughening. After annealing, the self-corrosion potential (table 20) decreased from -0.50V for bare iron to -0.55V for structured iron. The self-corrosion potential after modification underwent a marked increase to -0.27V and -0.23V for stearic acid and silane, respectively. The corresponding self-corrosion current densities were 0.12 and 0.083µA/cm². These values were much lower than the 0.54 and 1.44µA/cm² for bare and structured iron, respectively.

Table 20. Corrosion parameters of iron in 3.5wt%NaCl solution

Surface	E_{corr} (V_{SCE})	I_{corr} ($\mu A/cm^2$)
bare	-0.50	0.54
structured	-0.55	1.44
stearic acid modified	-0.27	0.12
silane modified	-0.23	0.083

A process based upon K_2CO_3 was used to create 3-dimensional flower-like Fe_3O_4 micro-nano flakes on the surface of iron via *in situ* hydrothermal synthesis[90]. The width of the nano-flakes ranged from 50 to 100nm, with a length of about 3µm. The morphology of the Fe_3O_4 nanostructures could be varied from simple isolated nano-flakes to ordered 3-dimensional flower-like shapes by increasing the reaction temperature. The wettability of surfaces with the latter flower-like micro-nano flakes was changed from hydrophilic to superhydrophobic by modification with vinyl tri-ethoxysilane. The static water contact-angles on the modified surfaces were greater than 150°, and this was attributed to the modification and to the hierarchical structure. The surfaces retained good superhydrophobic stability during long-term storage. Because the surface roughness played a key role, the effect of roughness upon wetting were studied by comparing the water contact-angles which resulted from various reaction temperatures, before and after the vinyl tri-ethoxysilane modification. When the reaction temperature was 120C, few flower-like micro-nano structures were formed. The surface had a contact-angle of 23° and a roughness of 114.2nm. When the temperature was increased to 180C, the flower-like structures grew bigger, contiguous and essentially covered the surface. The water contact-angle and roughness increased to 37° and 147.5nm. The increase from 23° to 37° was far from reaching the superhydrophobicity level. Following vinyl tri-ethoxysilane modification, the roughnesses of the surfaces obtained by 120C and 180C treatment were up to 155.1nm and 175.6nm, respectively, and the contact-angle was changed from 124° to 157°; indicating a change from hydrophobicity to superhydrophobicity. Surfaces which were obtained by 180C treatment had a sliding-angle of 1°. This all emphasized the importance of surface roughness in improving surface hydrophobicity. The superhydrophobicity of the modified surfaces was attributed to two factors. Before vinyl tri-ethoxysilane self-assembly, the terminal groups on the oxidized surface were hydrophilic oxygen atoms. Following complete vinyl tri-ethoxysilane self-assembly, the oxygen atoms were unavailable and the surface became hydrophobic because of the hydrophobic alkyl portion of the vinyl tri-ethoxysilane. Air was meanwhile trapped between the texture features and prevented water droplets from wetting the surface; again increasing hydrophobicity.

Chemical deposition was used to create micro-nano structures on iron sheet with the aid of candle soot as a surface template[91]. A soot coating was first deposited on the iron surface by placing the iron sheet over a candle-flame. Particles with a size of 10 to 100nm loosely accumulated and connected on the iron surface to form a fractal-like network that comprised micromeshes, sub-micromeshes and nano-meshes. The largest micro-meshes size could attain 5μm, and the larger micro-meshes contained small sub-micromeshes or nano-meshes which constituted a porous structure. The iron was then immersed in copper dichloride ethanol solution for 0.5h, yielding a copper layer having a 3-level structure comprising micro-protrusions, sub-micron papillae and nano-sheets. The size of the micro-protrusions ranged from 2μ to 8μm. The large micro-protrusions comprised sub-micropapillae which in turn were made up of nano-sheet structures with a thickness of 10 to 50nm. The iron was finally modified with octadecanoic acid. The resultant copper surface on the iron sheet was very superhydrophobic, with water contact-angles and sliding-angles attaining 156.9° and 4.6°, respectively.

Hierarchical structures were created on 2N5-purity iron by abrasion, calcination and modification, and the results showed that superhydrophobicity and adhesion depend upon micro-nano scale surface patterns[92]. The degree of adhesion could be controlled by adjusting the abrasion process so as to obtain patterns having an optimum ratio of height to width. Material was abraded manually with sandpaper so as form micro-scale patterns. Calcination was carried out in air at 400C for 4h. Modification was carried out via immersion or vaporization, with the former being used to incorporate 1H,1H,2H,2H-perfluorodecyltri-ethoxysilane, octadecane thiol, oleic acid or sodium dodecyl sulphate and the latter being used to incorporate paraffin or lard. The resultant surfaces could be classified into superhydrophobic with high adhesion, superhydrophobic with moderate adhesion and superhydrophobic with low adhesion. The nano-scale patterns of high-adhesion and low-adhesion surfaces were not significantly different, but the micro-scale pattern of the former changed markedly to a relatively flat shallow structure with widths and depths of about 1μm. Typical values were a contact-angle of 158.3° and a hysteresis of 91.0°. The maximum adhesive force was of the order of 195.8μN. The pH-value had essentially no effect upon water-wettability or adhesion in the case of low-adhesion surfaces but, in the case of high-adhesion surfaces, pH-values of 2 and 1 led to contact-angles of 152° and 136°, respectively. The rolling-angle meanwhile decreased from 180° to 85° and 33°, respectively. It was surmised that high acidity destroyed the microstructure, given that the high-adhesion surface had no air-film in its pattern which could prevent direct contact between acid and surface. Between -10C and 200C, the low-adhesion surfaces maintained their contact-angle at 157° to 164° over the entire temperature range, while the rolling-angle decreased from 18° to 5° as the temperature increased from -10C to 30C. It then gradually decreased to 3° with increasing temperature. In the case of high-adhesion surfaces, the contact-angle increased from 150° at -10C, to 155° between 15C and 45C, and then decreased to less 150° at 70C. The rolling-angle remained at 180°, from -10C to 65C, and then decreased to 90° at 70C. This change was attributed to a transition from Wenzel behaviour to Cassie behaviour at higher temperatures due to the existence of a vapour film within the hierarchical structure which diminished contact with the rough structure. When exposed to atmospheric air, the low-adhesion surface retained superhydrophobicity for up to 120 days. The high-adhesion surface lost its superhydrophobicity after 80 days due to a high affinity for contaminants. In both cases, superhydrophobicity was recovered by ethanol-washing. When immersed in 1M NaCl solution, the low-adhesion surface retained its superhydrophobicity for 5 days, while the high-adhesion surface could do so for only 3 days. The differing corrosion behaviours were attributed to the differing abilities of the surface microstructures to retain air and isolate corrosive media from the metal surface. The rolling-angle of low-adhesion surfaces increased markedly with the number of tape-removal cycles and reached 180° after losing superhydrophobicity. Residual adhesive and torn nano-scale microstructures caused a loss of superhydrophobicity and increased adhesion. Superhydrophobicity could be obtained for all sandpapers which were used.

For a given sandpaper mesh-size, the contact-angle increased with increasing degree of abrasion. The highest contact-angle, 165°, was found for the longest abrasion times and a mesh-size of 600 (particle-size of 10 to 14μm). It was noted that this corresponded to the size of the micro-scale pattern of a lotus-leaf or rose-petal. When the mesh-size was too large or too small, more abrasion was required in order to impart superhydrophobicity. The contact-angle increased significantly with increasing calcination temperature. The abraded iron surface became superhydrophobic for a calcination temperature of 300C but the rolling-angle decreased to 2° below and above 400C. Increasing the temperature to 600C caused the Fe_3O_4 layer to detach from the iron. Below 250C, there was no nanostructure and the contact-angle was 135.9°. At below 300C, scattered Fe_2O_3 nano-needles formed, leading to a contact-angle of 156.2°. Upon increasing the calcination temperature to 400C, the nanostructures became Fe_3O_4 nano-flakes which were larger, and more densely distributed. This increased the contact-angle to 160.1°. Calcination at 450C produced even larger nano-flakes and led to a contact-angle of 161.0°. It was noted that nanostructure provided enough roughness for superhydrophobicity and enough air for low adhesion. With increasing heat-treatment time, more and more Fe_3O_4 nano-flakes were generated and the contact-angle increased from 146.8° at 0h to 159.7° at 2h, before remaining at 161°. The rolling-angle meanwhile decreased from 180° at 0h to 2° at 2h.

Table 21. Roughness and contact-angle of etched mild-steel surfaces

Etching (min)	Initial Roughness (μm)	Final Roughness (μm)	Angle (°)
2	0.267	1.797	141
4	0.280	2.430	147.67
6	0.277	2.980	150.67
8	0.260	3.717	150.67
10	0.270	4.213	154

Superhydrophobic surfaces were created on mild steel surface by increasing the roughness (table 21) and reducing the surface energy[93]. The steel was etched with hydrochloric acid solution for 600s, and then immersed in stearic acid solution for 3h. The resultant surfaces had the usual combination of nano and micro features and the roughness of the as-prepared surface was 4.213μm and had a pine-cone structure. This hierarchical structure comprised numerous grooves within which air could be trapped and lead to a larger contact-angle. The latter could be up to 154°, with a lowered hysteresis. The as-prepared superhydrophobic surfaces also self-cleaned themselves of dirt as water droplets rolled over the surface. Samples were tilted at an angle of 35° and dusted (25mg/cm^2) with fine red sand. Water droplets (4µl) were dropped from a height of 5mm. The dust was washed off by 50 drops. A freezing delay-time of 392s was also observed.

Table 22. Corrosion parameters for mild steel in simulated seawater

Surface	E_{corr} (mV$_{SCE}$)	I_{corr} (A/cm^2)
bare	-600.2	1.5 x 10^{-5}
nickel coated	-208.1	1.5 x 10^{-7}
nickel-graphene coated	-150.3	2.3 x 10^{-8}

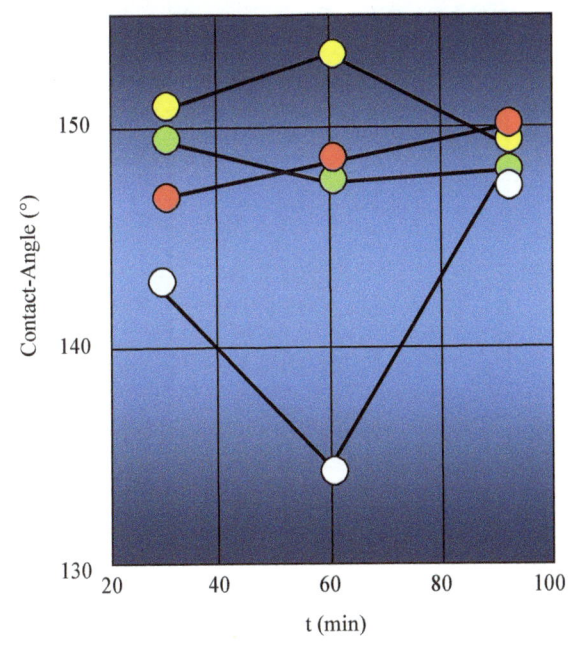

Figure 6. Contact-angle on mild steel as a function of annealing time for various plating conditions. Yellow: 6A/dm^2, green: 5A/dm^2, red: 4A/dm^2, white: 3A/dm^2

Superhydrophobic nickel/graphene coatings were deposited on mild steel via 1-step electrodeposition[94]. Coatings which were deposited using a graphene-oxide concentration of 0.2g/l and a current density of 4A/dm² had dual-roughness structure. When the graphene-oxide concentration was 0.3g/l, the surface had nodular bulges and the contact-angle decreased to 148.3°. The impairment of the superhydrophobicity was attributed to the agglomeration of graphene, and the lower graphene-oxide concentration was concluded to be the optimum one. It produced surfaces having a low adhesion to droplets. Following storage (14 days, vacuum), the coating had a contact-angle of 156.1° and a sliding-angle of 6.2°, while the corrosion current-density (table 22) was 2.3 x 10⁻⁸A/cm².

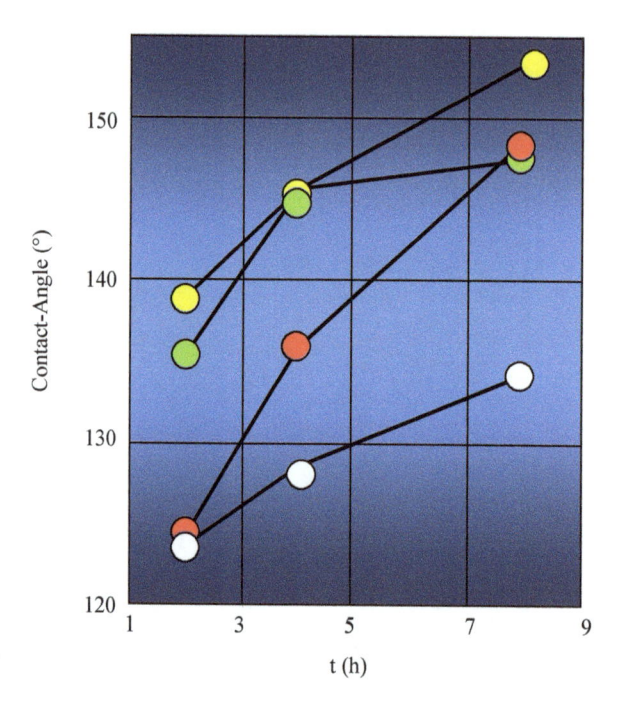

Figure 7. Contact-angle on mild steel as a function of plating time for various conditions. Yellow: 6A/dm², green: 5A/dm², red: 4A/dm², white: 3A/dm²

Electrochemical machining using a mixed aqueous solution of Na_2MoO_4 and NaCl was used to create superhydrophobic iron surfaces (pure [DT4E], mild steel [Q235], medium-carbon steel [C45]) via fluoro-alkylsilane modification[95]. This method rapidly generated rough microstructures at low electrical settings, and recycled the corrosion products. The morphology was hierarchical, with bell-shaped protrusions. Before modification, water droplets rapidly spread on the prepared film and the contact-angle was essentially zero, indicating superhydrophilicity. After modification, the surface had a water contact-angle of 167.2° and a sliding-angle of about 2°. The superhydrophobic surface exhibited abrasion-resistance plus self-cleaning and anti-icing and anti-fogging capabilities.

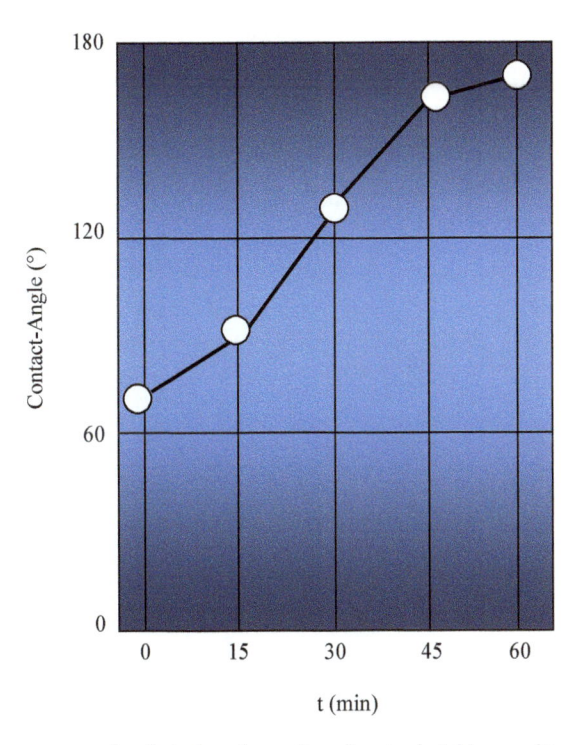

Figure 8. Water contact-angle of steel surface after chemical etching and treatment with lauric acid. The etching time was varied from 0.25h to 1h, while the treatment time with lauric acid was kept constant

A method which was based upon electrodeposition and annealing created superhydrophobic surfaces on mild steel without the use of chemical modification[96]. Plating parameters such as the current and time markedly affected the surface morphology. At a current density of 6A/dm² (figures 6 and 7), increasing the deposition time increased the protrusion-size. Increases in protrusions led to an increase in the percentage of solid/liquid contact-area. The optimum parameters yielded a rough surface with a multi-level structure, plus copper oxides, zinc oxides and many hydrocarbons. Transition-metal oxides which were on the surface adsorbed hydrocarbons from the air during annealing process, and that lowered the surface energy. The corrosion behaviour was evaluated in 3.5wt%NaCl solution by means of alternating-current impedance spectroscopy, showing that the superhydrophobic surface led to a great improvement in corrosion resistance.

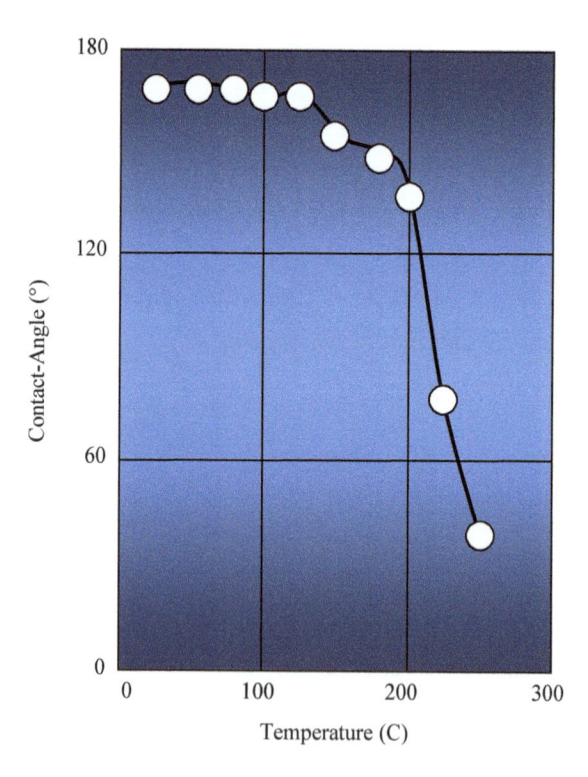

Figure 9. Contact-angle on superhydrophobic steel as a function of annealing temperature

Superhydrophobic coatings were prepared on steel surface by etching with a mixture of hydrochloric and nitric acids, followed by treatment with lauric acid[97]. Before treatment the clean steel surface was hydrophilic, with a static water contact-angle of 72°. The etching-time was varied from 15 to 60min, while the lauric acid treatment-time remained constant. By optimizing the etching conditions, a static water contact-angle of 170° and a tilt-angle of 5° was obtained. During etching, the shape and size of the microstructures changed and affected the contact-angle. After 15min of etching and 24h of lauric-acid treatment, the contact-angle had increased from 72° to about 90°. This increase was attributed to the lauric acid alone, and 15min of etching was insufficient to produce enough roughness. With increasing etching-time, the contact-angle increased. Following 30min of etching and 24h of treatment with lauric acid, the contact-angle was 130°. The angle further increased to 163° when the etching-time was 45min. The associated sliding-angle was greater than 10°, and further etching was deemed necessary (figures 8 and 9). Following 60min of etching and 24h of treatment with lauric acid, the surface was superhydrophobic, with a contact-angle of 170° and a sliding-angle of 5°. The thermal stability of the coating was tested by annealing (25 to 250C, 1h). The contact and tilt angles remain constant after annealing at 25 to 125C, indicating thermal stability. Upon annealing at 150C and 175C, the contact-angle decreased to 155° and 150°, respectively. The tilt-angle increased rapidly upon annealing at 150 to 175C, indicating so-called sticky superhydrophobicity. After annealing at 200C or above, the surface was no longer superhydrophobic, due to damage to the coating. The surfaces were also tested by immersion in 3.5%NaCl solution with a pH-value of 8. The coating was unaffected by up to 6h of immersion. At longer times, the contact-angle decreased with immersion time, again due to damage to the surface. By annealing at 300C for 6h, the surface was deliberately damaged and the contact-angle was less than 10°. In order to restore superhydrophobicity, the damaged material was immersed in an ethanoic solution of lauric acid for 24h. The contact and tilt angles were then about 165° and 5°, respectively. The corrosion-resistance of the superhydrophobic surface was tested by immersion in 3.5wt%NaCl for 30min, and compared with that of the uncoated steel. The corrosion potential and corrosion current density (table 23) were obtained from potentiodynamic polarization curves. The I_{corr} value after coating decreased and the E_{corr} increased. Ions and water molecules in the solution immediately reacted with the uncoated surface, due to its hydrophilic character and polar–polar attraction. Coating made the surface superhydrophobic and a non-polar barrier. Contact between the ions and steel surface was also very low, due to the air trapped in the grooves on the coated steel surface. The corrosion-resistance was thus improved by coating.

Table 23. Corrosion potential and corrosion current density for steel surfaces in 3.5%NaCl solution

Surface	E_{corr} (V)	I_{corr} (mA/cm^2)
bare	0.131	0.008
coated	0.169	0.002

A mixture of antiformin solution and hydrogen peroxide was used to grow porous structures on steel, which had a lotus-leaf like hierarchical micro-nano form following ultrasonic treatment[98]. Superhydrophobic surfaces having such micro-nano structures tend to be mechanically weak. The modified superhydrophobic surface could withstand sandpaper abrasion over a distance of 2.24m under a pressure of 24.50kPa without losing superhydrophobicity. The water contact-angles were between 150° and 165°, and the sliding-angles ranged from 5° to 9° during abrasion cycles. The as-prepared superhydrophobic surface offered excellent self-cleaning and anti-corrosion capabilities. The anti-corrosion capability of the superhydrophobic surface was improved, as seen from the lower I_{corr} (12.706mm/cm^2) and higher E_{corr} (-0.511V) values as compared with those of the bare material.

AISI1018

Superhydrophobic surfaces were prepared on AISI1018 low-carbon steel by means of laser-texturing using a wavelength of 1064nm, a frequency of 50kHz and a scanning speed of 100mm/s[99]. Circular patterns with a diameter and pitch of 200μm were created. Wax and candle-soot were then used to decrease the surface energy. The water contact-angle (10μl droplet) was thereby increased, from the 87° of the untreated surface, to a value of 155.6°. Droplets easily rolled-off from surfaces tilted at about 2°.

C45

Superhydrophobic textured surfaces were prepared on S45C steel by firstly grinding and polishing and then etching in hydrofluoric acid and hydrogen peroxide mixtures for 30s at room temperature[100]. The textured surface had island-like protrusions, micro-pits and nano-flakes. These multi-scale structures, plus low surface-energy molecules, created superhydrophobic surfaces with a contact-angle of 158° and a sliding-angle of 3°. Potentiodynamic polarization tests showed that the as-prepared superhydrophobic surface offered excellent corrosion-resistance. Tribological tests showed that the friction-coefficient of the superhydrophobic surface was 0.11 for 6000s and the superhydrophobicity did not decrease during abrasion tests.

Superhydrophobic surfaces with a micro-nano papilla structure were prepared on C45 steel by using a 1-step electrochemical method[101]. The initial surface of the steel was hydrophilic, with a contact-angle of about 50°, and droplets adhered even when the surface was inverted. The treated surfaces had a water contact-angle of 160.5° and a sliding angle of 2°. The chemical composition of the surface was that of an iron complex with an organic acid. Electrochemical measurements showed that the superhydrophobic surface greatly improved the corrosion resistance of the steel in 3.5wt%NaCl solution: the layer provided a stable air/liquid interface which barred penetration of the corrosive medium.

Superhydrophobic surfaces with hierarchical micro-nano papillae structures were created[102] on C45 steel by using a 1-step electrochemical method, and led to a water contact-angle of 160.5° and a sliding-angle of 2°. The composition of the surface film was an iron complex with an organic acid. Electrochemical measurements showed that the superhydrophobic surface greatly improved the corrosion resistance of the carbon steel in 3.5wt%NaCl solution. The superhydrophobic layer acted as a barrier and a stable air/liquid interface inhibited the penetration of the corrosive medium. The as-prepared steel offered an excellent self-cleaning capability.

Table 24. Corrosion parameters of Q235 steel in 3.5wt%NaCl solution

Surface	E_{corr} (mV)	I_{corr} (A/cm^2)
bare	1045	3.940×10^{-5}
as-deposited	-423	2.982×10^{-6}
modified	-276	8.008×10^{-8}

Q235

Amorphous alloys make good hydrophobic coatings due to their low surface energy. Detonation-spraying and surface-modification were used to create hard and dense superhydrophobic iron-based amorphous coatings on Q235 steel[103]. The water contact-angle of the as-deposited coating was 81.0°, as compared with the 74.4° of the Q235 substrate. The lower surface energy of the as-deposited coating made it easier to create the hydrophobic surface. The water contact-angle was 160.8° following fluoroalkylsilane modification. The corrosion current-density (table 24) of the superhydrophobic coating was 2 orders-of-magnitude lower than that of the as-deposited coating: $8.008 \times 10^{-8} A/cm^2$ as compared with $5.473 \times 10^{-6} A/cm^2$. The corrosion potential of the superhydrophobic coating meanwhile shifted by 34mV to the positive side, as compared with that (-310mV) of the as-deposited coating. Due to the selective corrosion in HCl/H_2O_2 solution, surfaces having a uniform needle-like micro-nano structure were formed. The hydrophobicity of the surface did not exhibit a linear dependence upon the surface roughness.

Superhydrophobic surfaces on Q235 carbon steel, with a water contact-angle of 161.6° and a contact-angle hysteresis of 0.8°, were prepared[104]. The as-prepared surfaces repelled solutions with pH-value ranging from 1 to 14; hydrochloric acid solution, with a pH of 1, maintained a spheroidal shape on the as-prepared superhydrophobic surface at room temperature. The polished steel was very smooth, with an average roughness of 39nm. It was immersed in H_2O_2-H_2SO_4 mixtures in order to reveal the microstructures, and the structured steel had a rough surface bearing complex and regular needle-like forms. The latter were some 200nm long and 10nm wide; resulting in nano-pores with a diameter of 100nm. The roughness markedly increased to 491nm, while the grafting of 1H,1H,2H,2H-perfluorodecyltri-ethoxysilane to the surface reduced the roughness to 423nm. The topography of the structured surface following this modification was similar to that of the structured steel surface. These microstructures accorded enough space to entrap an air layer. The polished surface contained just iron, while the structured steel surface contained iron, oxygen and carbon. Following modification, the surface contained iron, fluorine and silicon; indicating successful grafting. The water contact-angles of polished surfaces and of modified polished surfaces were 90.3° and 99.1°, respectively. The modified structured sample had the above-mentioned contact-angle and hysteresis, and that combination of properties indicated that the modified surface was in the Cassie-Baxter state. The area between air and liquid on the surface of the modified sample was much larger than the contact area between the solid and liquid, indicating that the surface was superhydrophobic. Potentiodynamic polarization studies of the surface yielded lower I_{corr} values and higher E_{corr} values, indicating a higher corrosion-resistance. The I_{corr} of superhydrophobic Q235 steel was $0.577 \mu A/cm^2$, and this was much lower than that for the untreated steel or the structured steel, 6.601 and $3.917 \mu A/cm^2$, respectively. The E_{corr} value of the superhydrophobic surface meanwhile changed positively, as compared with other surfaces. The overall results suggested that the superhydrophobic steel offered an excellent corrosion resistance. As a test, highly corrosive droplets of HCl-$CuSO_4$ with a pH-value of 1 were dropped onto untreated, structured and superhydrophobic surfaces. The untreated surface began to corrode after 10s and the corroded area gradually increased over 60s. The structured surface offered a certain corrosion-resistance, but damage occurred after 420s.

Table 25. Corrosion-rate of St3 steel, with the 4th type of superhydrophobic coating, in NACE + 400mg/l H$_2$S solution

Surface	Exposure Time (h)	Corrosion-Rate (g/m^{2}h)
bare	24	0.522
bare	240	0.201
coated	24	0.072
coated	240	0.070

Steel surfaces were prepared, the superhydrophobicity of which could survive severe bending. The roughness on the steel was created by etching in acid, and the surface energy was lowered by silane treatment[105]. The optimum etching time in sulphuric acid was 8h. The water contact-angle was 164° and the sliding-angle was 9° following modification using methyltrichlorosilane. The steel could be bent through 90° and 180° without the wetting properties in the bent area exhibiting any change in superhydrophobicity. This surface also exhibited excellent self-cleaning, as well as retaining superhydrophobic wetting under a jet of water.

Table 26. Corrosion-rate and corrosion potential of 3rd type of superhydrophobic coating on St3 steel in NACE + 400mg/l H$_2$S

Surface	Exposure Time (h)	E_{corr} (V)	I_{corr} (A/m^2)
bare	0.25	-0.47	0.29
bare	24	-0.57	0.20
bare	48	-0.61	0.25
bare	96	-0.57	0.24
bare	120	-0.54	0.25
bare	144	-0.48	0.23

bare	168	-0.42	0.24
coated	0.25	-0.47	0.10
coated	24	-0.48	0.03
coated	48	-0.47	0.028
coated	96	-0.43	0.025
coated	120	-0.50	0.04
coated	144	-0.52	0.05
coated	168	-0.50	0.10

GCr15

Etching of chromium bearing steel, GCr15, with H_2SO_4 and H_2O_2 was used to produce a superhydrophobic surface with a water contact-angle of 163.5° and a sliding-angle of essentially zero, following modification with 1H,1H,2H,2H-perfluoro-alkyltri-ethoxysilane[106]. By means of perfluoropolyether infusion, a slippery liquid-infused porous surface was created which had a water contact-angle of 115.6° and a sliding-angle of 2.27°. The material could retain superhydrophobicity after traversing 100cm of 1000-grit sandpaper under a pressure of 100g. A good corrosion resistance was signalled by increased positive corrosion potentials and lower corrosion current densities in contact with various solutions. Even when the superhydrophobic and slippery properties were impaired during long-term soaking in saline solution, the steel could regain its capabilities during re-treatment.

Table 27. Corrosion parameters for stainless steel in 0.5M NaCl

Surface	E_{corr} (mV)	I_{corr} (μA/cm^2)
bare	-455.03	0.0710569
Ni/stearic-acid	-434.56	0.0041915
Ni/stearic-acid/metal-organic-framework	-392.42	0.0007017

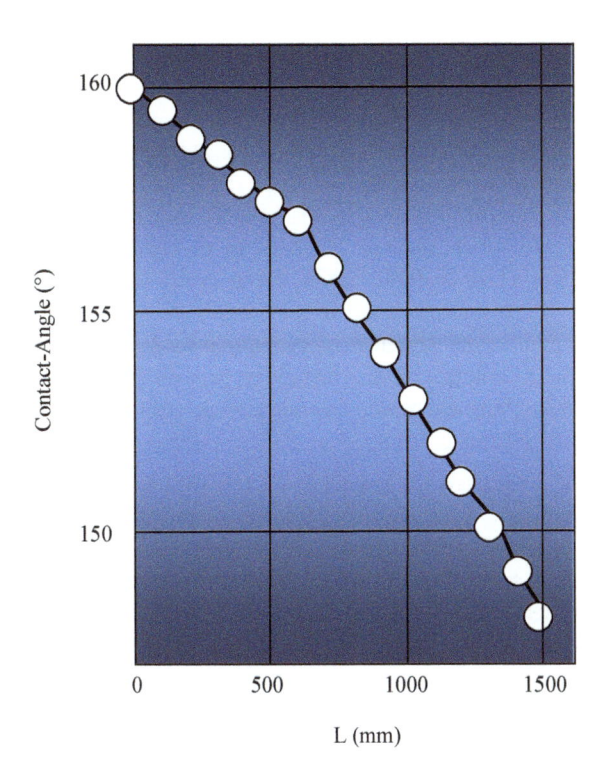

Figure 10. Contact-angle as a function of abrasion length for Ni/stearic-acid coated superhydrophobic stainless steel

St3

Four types of nano-composite superhydrophobic coating were applied to St3 steel[107]. The first type of coating involved texturing the surface by using nanosecond infra-red laser (pulse duration 50ns, pulse frequency 20kHz, peak pulse power 0.95mJ, scanning rate 50mm/s, scanning density 150lines/mm) radiation and then modifying it via the chemisorption of methoxy{3-[(2,2,3,3,4,4,5,5,6,6,7,7,8,8,8-pentadecafluoro-octyl)oxy]propyl}silane. The second type of coating was a nano-sized composite layer consisting of Aerosil nanoparticle aggregates, coated with the above hydrophobic silane which was added to the first coating. The third type of coating involved texturing using a nanosecond infra-red laser, and chemisorbing $CF_3(CF_2)_6(CH_2)O\text{-}(CH_2)_2C(OCH_3)_3$ from the vapour phase. The fourth coating involved texturing the surface with a nanosecond infra-red laser, followed by treatment with ultra-violet radiation in the presence of ozone plasma. Samples were the kept in a closed container with saturated vapours of the above silane. The contact-angles on the various surfaces ranged from 168 to 172°. Tests were carried out in a neutral chloride medium, in NACE (5g/l NaCl, 0.25g/l CH$_3$COOH, pH = 3.6) + 400mg/l H$_2$S and in 100% humidity. In the latter case, superhydrophobic coatings of the first 2 types did not change over 112 days. In the neutral chloride medium, their protective effect was close to 96%. Environments which contained sulphide led to fragility of the anti-corrosion properties of the superhydrophobic coatings. The use of ultra-violet irradiation in the presence of ozone plasma led to an increase in anti-corrosion properties. Coatings of the first 3 types exhibited highly anti-corrosive behaviour in 100% humidity and neutral chloride solution (tables 25 and 26).

Superhydrophobic stainless-steel surfaces were created by using biological metal-organic frameworks[108]. The latter were synthesized by using aspartic acid as a linker and copper ions as the core metal. Two types of coatings were prepared by means of the electrodeposition of nickel, which incorporated a metal-organic framework, followed by soaking in an ethanoic solution of stearic acid. The water contact-angles of the Ni/stearic-acid and Ni/stearic-acid/metal-organic framework surfaces were 160° and 168°, respectively. The coating was chemically and mechanically stable. The Ni/stearic-acid coating maintained its superhydrophobicity for pH-values ranging from 3 to 11. The coating with the metal-organic framework retained superhydrophobicity for pH-values ranging from 1 to 13. The Ni/stearic-acid coating retained superhydrophobicity after abrasion over lengths of up to 1300mm. The coating with the metal-organic framework content retained superhydrophobicity after abrasion over a length of 2700mm (figures 10 and 11). Both coatings imparted a markedly improved corrosion protection in 0.5M NaCl solution, compared with bare stainless steel, with protection efficiencies of some 94% and 99%, respectively. The corrosion current density (table 27) for the steel coated without the metal-organic framework was 0.0041915µA/cm^2, and that for the coating with the metal-organic framework was 0.0041915µA/cm^2.

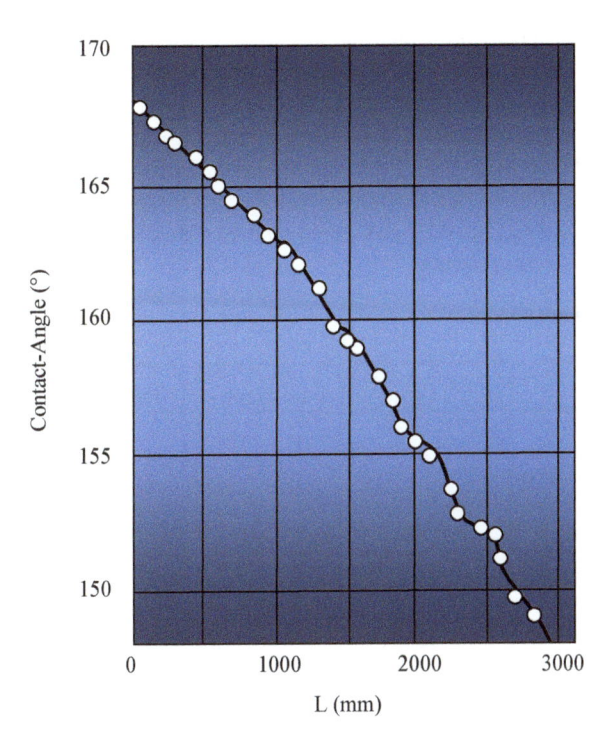

Figure 11. Contact-angle as a function of abrasion length for Ni/stearic-acid/metal-organic-framework coated superhydrophobic stainless steel

30Cr2NiWVA

Superhydrophobic surfaces were produced on 30Cr2NiWVA steel by electrochemical etching (25wt%$NaNO_3$, 5A/cm^2, 60s) and fluoro-alkylsilane modification[109]. Following the etching, the surface was covered by a rough coral-like film with features which were 10 to 20μm long and 1 to 2μm wide. properties. The surface retained its superhydrophobicity after exposure to air for two years, with the contact-angle before and after exposure being 165° and 154°, respectively. An untreated steel and the superhydrophobic steel were tilted at 10° in an environment at 16.5C. Water easily wetted the untreated surface, ice gradually formed with the passing of time. Water did not wet the superhydrophobic surface, and no ice formed, leading to an excellent anti-icing behaviour. When the superhydrophobic surfaces were exposed to 3.5wt%NaCl solution, the corrosion-potential and corrosion current-density were -0.375V and 3.16×10^{-8} A/cm^2, respectively, as compared with -0.47V and 1.78×10^{-7} A/cm^2 for untreated surfaces. This increase in the corrosion-potential and decrease in the corrosion current-density meant that the treated surface protected the steel from corrosion. Surfaces were tilted at 45°, and water droplets (15.2µl) were used to remove chalk dust at a flow-rate of 1.46ml/min. The contaminant was dissolved by water and stayed on the untreated surface, but was picked up and rolled off the treated surface.

Table 28. Corrosion parameters for 17-4 PH stainless steel in saline solution

Surface	E_{corr} ($V_{Ag/AgCl}$)	I_{corr} (A/cm^2)	Corrosion Rate (mm/y)
bare	-0.136	1.035×10^{-6}	0.012
coated	-0.110	4.018×10^{-7}	0.005
coated channelled	-0.209	2.776×10^{-6}	0.032

17-4

These treatments are not always entirely successful. Nanosecond fibre laser texturing of 17-4PH Cr-Ni-Cu stainless steel, followed by addition of a low-energy coating, produced a surface with steady-state contact-angles of up to 145°. The surface was subjected to abrasion with 600-grit silicon carbide paper before the texturing, and the laser wavelength was 1060nm, the laser power was 12W, the frequency was 20kHz, the scanning speed was 600mm/s and the pulse width was 60ns. Non-textured surfaces had a contact-angle of 121°. The laser-texturing did not affect the microstructure of the base metal[110], nor did it cause material loss during the process. The resultant higher corrosion current density, lower corrosion potential and higher corrosion-rate of the textured surfaces were attributed to the size of the micro-grooves, which could not retain entrapped air within the hierarchical structure when immersed in a corrosive medium. The hydrophobic coating consisted of 2-(difluoromethoxymethyl)-1,1,1,2,3,3,3-heptafluoropropane and 4-methoxy-1,1,1,2,2,3,3,4,4-nona-fluorobutan. The combined effects of the micro-roughness and organic coating produced a hydrophobic surface with contact-angles which were just below the required contact-angle for superhydrophobicity. The corrosion-rate of the base metal exhibited a decrease from 0.012 to 0.005mm/year after applying the hydrophobic coating. This essentially 50% increase in corrosion-resistance was attributed to the protective nature of the coating, but the corrosion-rate of a coated channelled surface increased to 0.032mm/year, even though the coated textured surfaces were more hydrophobic. The corrosion current density of the hydrophobic laser-textured surfaces followed the same trend as the corrosion-rate data (table 28) because I_{corr} was directly related to the corrosion-rate. On the other hand, the E_{corr} values exhibited a marked shift to less noble for laser-textured surfaces; indicating a higher corrosion tendency. So the lowest corrosion potential with the highest corrosion current density, leading to the highest corrosion-rate, was found for a laser-textured hydrophobic sample. The water-repelling property of this stainless steel was therefore ineffective in preventing chloride ions from approaching the substrate. This was attributed to the size of the micro-grooves on the surface, which were incapable of retaining entrapped air.

AISI304

Superhydrophobic surfaces were prepared on AISI304 austenitic stainless steel[111] by using a 2-step method involving etching in HF solution, followed by fluorination, leading to a maximum water contact-angle of 166° and a sliding-angle of 5°. In the second step, the surfaces were immersed in 0.1wt%NaCl solution at 100C. The water contact-angle had then increased to 168° and the sliding-angle had decreased to 2°. The good superhydrophobicity was retained after 1 month of storage in air and water.

AISI316

Microstructures of various scales were created on 316L stainless steel by laser processing[112]. By varying the laser fluence (2.69, 3.96, 6.28, 8.14, 9.55J/cm^2), various sizes of micro-cracks and brain-like microstructures were obtained. The best static contact-angle of 160° and sliding-angle of 3°, were obtained when using a scanning interval of 30.0μm and a laser fluence of 8.14 J/cm^2. The laser-textured superhydrophobic surface markedly increased the corrosion resistance, and imparted an excellent self-cleaning capability.

X90

A superhydrophobic surface was prepared on X90 manganese pipeline steel by creating an hierarchical structure with nano-flakes via electrodeposition and immersion[113]. This led to a water contact-angle of about 157° and a sliding-angle of about 3° following fluorination modification. The surface offered efficient self-cleaning and anti-corrosion. The $CaCO_3$ crystals on the superhydrophobic coating were largely needle-like, unlike the rhombohedral $CaCO_3$ crystals on the substrate. This coating exhibited good long-term stability in air, and mechanical and thermal stability.

Magnesium

A flower-like superhydrophobic film was produced on pure magnesium by etching in H_2SO_4 and H_2O_2, followed by immersion in stearic acid ethanol solution[114]. The surface had a static water contact-angle of 154° and a sliding-angle of about 3°. The flower-like structure, and bonding of the stearic acid to the magnesium surface were deemed to be the cause of a good water-repellence. The transfer-resistance of the superhydrophobic surface was increased by some 4 times, over that of bare magnesium, after a 1h immersion in 0.1mol/*l* NaCl solution.

A superhydrophobic surface was produced by electrodepositing Mg-Mn-Ce alloy in an ethanol solution which contained cerium nitrate and myristic acid[115]. The shortest electrodeposition time which was required in order to obtain a superhydrophobic surface was about 30s and the surface had a maximum contact-angle of 159.8° with a sliding-angle of less than 2°. The surface greatly improved the corrosion behaviour in 3.5wt% aqueous solutions of NaCl, Na_2SO_4, $NaClO_3$ and $NaNO_3$.

A superhydrophobic anti-corrosion coating was created on magnesium alloy by using a hydrothermal method[116]. The as-prepared surface had a water contact-angle of 161.7° and a sliding-angle of 4.8°. The corrosion-resistance was tested in 3.5wt%NaCl solution by means of electrochemical impedance spectroscopy. As compared with the bare surface, the corrosion current density of the superhydrophobic surface decreased by an order-of-magnitude, from $7.943 \times 10^{-5} A/cm^2$ to $7.943 \times 10^{-6} A/cm^2$, implying an enhancement of the corrosion-resistance of the magnesium substrate in NaCl solution.

AZ31

A stable superhydrophobic surface was produced on AZ31 alloy by combining electrodeposition and modification with stearic acid[117]. The as-prepared surface, with leaf-like clusters, had a static water contact-angle of 156.2° and a sliding-angle as small as 1.0°. The surface greatly improved the corrosion resistance of alloy in 3.5wt%NaCl solution, and was stable in high-saline and corrosive liquids. The superhydrophobicity was retained following mechanical abrasion over 900mm, and maintained a good corrosion resistance after abrasion over 1100mm.

Table 29. Corrosion data for AZ31 in 3.5wt%NaCl

Electrolyte	E_{corr} (V_{SCE})	I_{corr} (A/cm^2)
-	-1.542	7.02 x 10^{-5}
1	-1.471	1.07 x 10^{-6}
2	-1.484	1.66 x 10^{-6}
3	-1.525	5.80 x 10^{-8}
4	-1.517	3.65 x 10^{-7}

One-step electrodeposition was used to construct a superhydrophobic coating on AZ31 alloy by using magnesium nitrate and an ethanol solution of stearic acid as the electrolyte[118]. Four electrolytes, having differing ratios of stearic acid to magnesium nitrate, were used to clarify the effect of the electrolyte. The addition of magnesium nitrate increased hydrogen evolution during electrodeposition, and this was detrimental to electrodeposition. Samples which were prepared using a 10:1 molar ratio of stearic acid to magnesium nitrate offered the best corrosion resistance, with a corrosion circuit density of 3.74 x 10^{-8}A/cm^2. This was far lower than the current density for the bare magnesium alloy. A less dense coating was found for specimens prepared using electrolyte-1, and was not suitable for imparting good corrosion resistance. For the other electrolytes, the surface of the coating consisted of micro-spheres which were made up of nano-sheets. The size of the nano-sheets depended upon the electrolyte, as did the size of the micro-spheres. The coatings contained mainly carbon, oxygen and magnesium, with carbon predominating. This was attributed to the stearic acid. A small amount of MgO existed, due to the easy oxidation of magnesium. It was deduced that the main component of the coatings was magnesium stearate. The contact-angles of the electrolyte-2, electrolyte-3 and electrolyte-4 samples all exceeded 150°: 152.7°, 156.2° and 155.1°, respectively. When the coated samples were in contact with aqueous solution, the proportions of the surface which were in contact with the solution were 40%, 16.5%, 12.6% and 13.7%, respectively. The contact-angle of the electrolyte-1 sample was only 136.4°. The corrosion potentials of the 4 samples (table 29) were close to those of the open-circuit potential time curves. The corrosion potentials of the coated samples were more positive than that of AZ31, and the changes in the corrosion current density were more marked. The sample with the lowest corrosion current density has a corrosion current density had a value which was 3 orders-of-magnitude lower than that of AZ31.

Following soaking for one week in corrosive media, the contact-angles for electrolytes 2, 3 and 4 were still close to superhydrophobic levels, with angles of 150.3°, 150.5° and 151.3°, respectively.

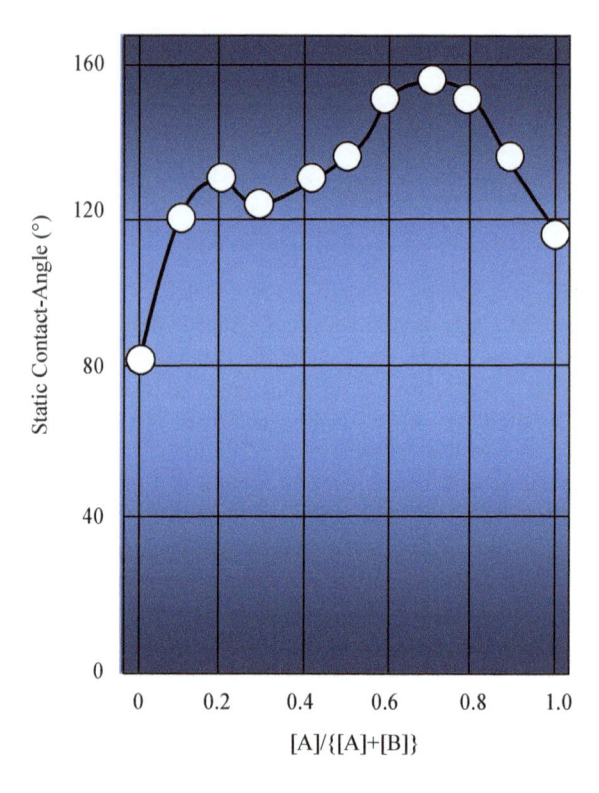

Figure 12. Static water contact-angles on AZ31 surfaces after immersion for 180s in solutions with various volume ratios of myristic acid (A) and cerium nitrate (B)

A superhydrophobic $Fe(OH)_3$ surface was obtained on AZ31 alloy via immersion, and modification in absolute stearic acid, using a hydrothermal method[119]. The surface had static water contact-angle of 163.7° and a water roll-off angle of less than 1°. It exhibited long-term durability in air, offered good chemical stability for pH-values ranging from 1 to 13 and was highly resistant to corrosion.

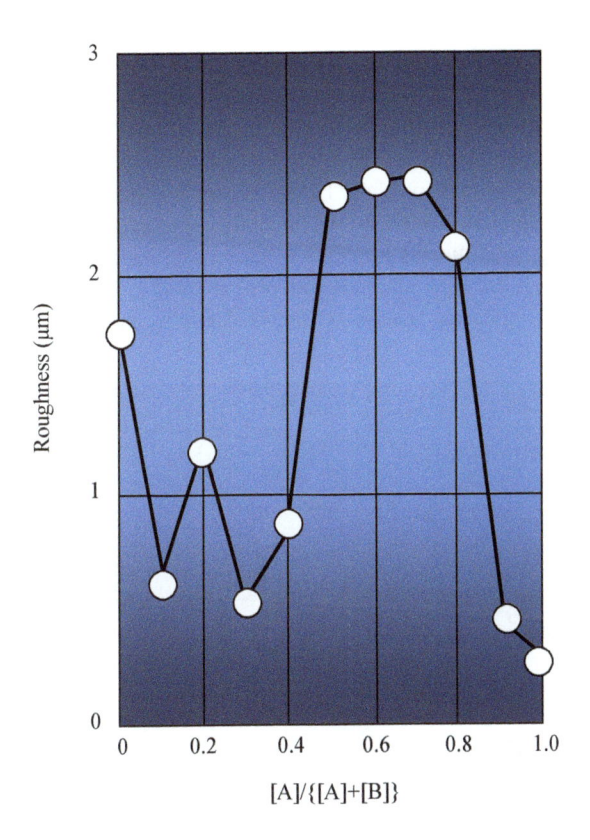

Figure 13. Surface-roughness of AZ31 after immersion for 180s in solutions with various volume ratios of myristic acid (A) and cerium nitrate (B)

Superhydrophobic surfaces were produced on AZ31 by means of 1-step immersion at room temperature[120]. Myristic acid-modified micro-nano surfaces had static water contact-angles of over 150° and a contact-angle hysteresis of less than 10°. The optimum volume-ratio of ethanoic myristic acid to aqueous cerium nitrate solutions for creating superhydrophobic surfaces was 0.7. One immersion solution (A) was prepared by introducing 5.03g of myristic acid into 100ml of ethanol. Another solution (B) was prepared by introducing 1.15g of cerium nitrate hexahydrate into 100ml of ultra-pure water. The solutions were used in volume-ratios of between 0:10 and 10:0, and the pH-values were all adjusted to 2.0. The effect of varying the solution-ratio upon contact-angle (figure 12) and surface-roughness (figure 13) was determined. The shortest treatment-time which was required in order to obtain superhydrophobic surfaces was 30s. It was shown that crystalline myristic acid could be formed on magnesium alloys by using a suitable molar ratio of cerium ions to myristic acid. The contact-angle hysteresis decreased with increasing immersion time. Potentiodynamic polarization measurements showed that the corrosion resistance of the immersion-treated AZ31 was markedly improved by the formation of superhydrophobic surfaces. The static water contact-angles after immersion for 12h in aqueous solutions with pH-values of 4, 7 or 10 were 90°, 119° and 138°, respectively; indicating that the chemical durability in a basic solution was high.

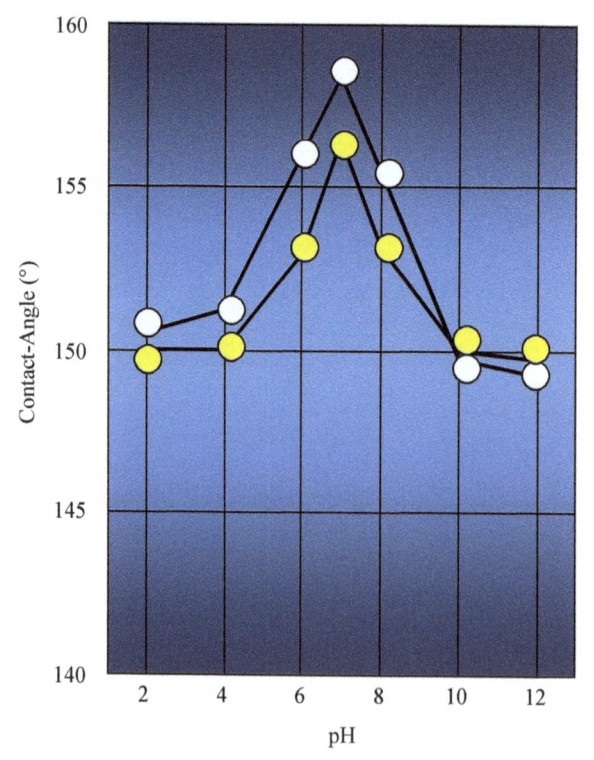

Figure 14. Relationship between contact-angle and pH-value of droplets on superhydrophobic AZ31. White: as-prepared, yellow: after one year in storage

A superhydrophobic surface was created on AZ31 by using a single-step hydrothermal method[121]. The as-prepared surface had a static water contact-angle of 156.7°, and this superhydrophobicity was maintained for more than one year (figure 14). It exhibited anti-icing behaviour in cold environments. The surface improved the corrosion resistance of the alloy substrate in 3.5wt%NaCl solution. The contact-angle of the just-abraded surface was 60°. Electrochemical parameters were derived from polarization curves for various immersion times (table 30). The I_{corr} value ($3.60 \times 10^{-6} A/cm^2$) of the superhydrophobic surface, when immersed for 3h, decreased by more than an order of magnitude when compared with that ($1.45 \times 10^{-4} A/cm^2$) of the untreated alloy when immersed for the same period. The I_{corr} value of the superhydrophobic surface after immersion in 3.5wt%NaCl aqueous solution for 6h was slightly shifted to approximately $5.50 \times 10^{-6} A/cm^2$. Following immersion for 9h, the I_{corr} value was $8.62 \times 10^{-6} A/cm^2$. Lower corrosion-current densities generally corresponded to lower corrosion rates and to an improved corrosion resistance. The increased corrosion was attributed to an air layer between the superhydrophobic surface and the NaCl aqueous solution. The thickness of the air layer remained relatively stable over 9h.

Table 30. Electrochemical parameters for AZ31 in 3.5wt%NaCl aqueous solution

Surface	Time (h)	E_{corr} ($V_{Ag/AgCl}$)	I_{corr} ($\mu A/cm^2$)
normal	3	-1.182	145
SHP	3	-1.169	3.60
SHP	6	-1.173	5.50
SHP	9	-1.180	8.62

AZ31B

A micro-nano structure superhydrophobic composite coating having self-healing and anti-corrosion properties was produced on AZ31B[122]. The composite was based upon lauric acid and hydrotalcite in a 3-layer arrangement with polydimethylsiloxane sandwiched between lauric acid/hydrotalcite powder and lauric acid/hydrotalcite film. The anti-corrosion performance of the as-prepared coatings (table 31) was determined using potentiodynamic polarization and electrochemical impedance spectroscopy. This showed that the E_{corr} of the bare substrate was the most negative. Following hydrothermal treatment, the hydrotalcite coating had a more positive E_{corr} value, but the value of I_{corr}

was close to that of the substrate. As usual, a lower I_{corr} or higher E_{corr} value presaged slower interface reaction and better anti-corrosion behaviour. The intercalated lauric acid and hydrotalcite coating had a much lower I_{corr} value than that of the bare substrate or hydrotalcite coating. This implied that the modification and intercalation of lauric acid could improve the corrosion resistance. The static contact-angle of the superhydrophobic coating was 155°. By trapping corrosive Cl⁻ ions and releasing protective lauric acid, the coating markedly reduced the corrosion rate. The air-film of the coating could also reduce contact with a corrosive liquid. Following heat treatment, the scratched coating could heal itself. This was because of the migration of C-H chains with low surface energy, provided by the polydimethylsiloxane. When the surface was scratched, the interior of the coating was exposed, leading to an increase in the surface free energy of the coating surface. During heat treatment, the molecules of the polydimethylsiloxane in the sub-surface became mobile. The C-H chains then migrated to, and accumulated at, the uppermost layer, thus healing the coating.

Table 31. Electrochemical data for coated AZ31B surfaces

Surface	E_{corr} (V)	I_{corr} (A/cm^2)
bare	-1.465	1.834×10^{-5}
hydrotalcite	-1.375	2.450×10^{-5}
intercalated lauric acid and hydrotalcite	-1.344	5.717×10^{-6}
lauric acid, hydrotalcite and polydimethylsiloxane	-1.339	1.065×10^{-7}
superhydrophobic	-1.307	6.447×10^{-9}

AZ61

A superhydrophobic film was created on electroless Ni-P plated AZ61 by using a hydrothermal method and was then immersed in stearic acid solution in order to improve the anti-corrosion and self-cleaning capabilities[123]. The plated Ni-P surface was hydrophilic, but became superhydrophobic after the further processing. The temperatures and times of the hydrothermal treatment affected the morphology and wettability of the surface. Petal-shaped nano-sheets, which resulted from using higher temperatures, imparted a higher hydrophobicity than did a lemon-grass like nano-structures which were produced by lower temperatures. The greatest contact-angle, 155.6°, and a sliding-angle of about 2°, were obtained by using a reaction-temperature of 120C and a reaction-time of 15h. Potentiodynamic polarization and electrochemical impedance spectroscopy characterized the corrosion resistance of the AZ61 substrate in 3.5%NaCl. For pH-values ranging from 2 to 12, the contact-angle ranged from 150° to 156°. The contact-angles remained higher than 145° in strongly acidic and strongly alkaline environments. For NaCl concentrations ranging from 1 to 5mol/l, the contact-angles were all greater than 150°. The superhydrophobicity thus survived strongly acidic, strongly alkaline and highly saline environments. The contact-angles were also maintained at above 150° during 180 days of air-exposure. The mechanical stability was tested by abrasion, with the superhydrophobic surface being rubbed across 400-grit sandpaper under a pressure of 9.8kPa, showing that the surface retained a contact-angle greater than 150° after travelling 120cm. This value fell to 145° after travelling 192cm. The superhydrophobic surface had an E_{corr} value of -139.3mV, a corrosion-inhibition efficiency of 99.8% and an I_{corr} value of $7.2 \times 10^{-2} \mu A/cm^2$ (table 32).

Table 32. Electrochemical corrosion data for AZ61 in 3.5%NaCl

Surface	E_{corr} (mV)	I_{corr} (mA/cm^2)
original	-1548.0	35.3
Ni-P coated	-656.7	19.8
superhydrophobic	-139.3	7.2×10^{-2}

Table 33. Corrosion data for AZ91 in 3.5%NaCl solution

Contact-Angle (°)	E_{corr} (V_{SCE})	I_{corr} (A/cm^2)
35	-1.48	1.66×10^{-3}
85	-1.46	8.69×10^{-4}
126	-1.45	5.84×10^{-4}
154	-1.42	4.99×10^{-4}

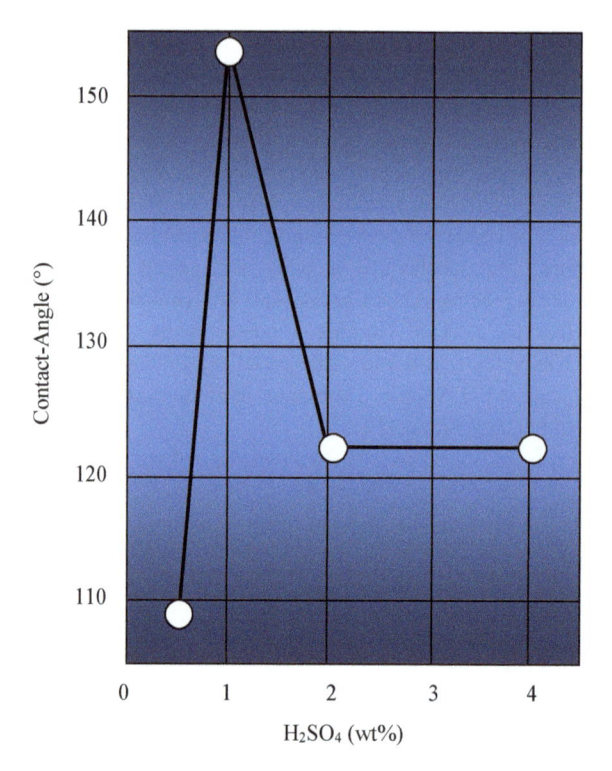

Figure 15. Effect of sulphuric acid concentration upon the water contact-angle of AZ91 surfaces

AZ91

A superhydrophobic coating on AZ91 was created sulphuric acid etching, $AgNO_3$ treatment and dodecyl mercaptan modification[124]. Optimum conditions were identified, and a surface with a water-contact angle of 154° and a sliding-angle of 5° was produced by using 1.0wt% of sulphuric acid solution for 240s, 0.001mol/l of $AgNO_3$ solution for 300s and 0.5mol/l of dodecyl mercaptan solution for 5h. The acid etching ensured a roughened surface, while the nitrate treatment resulted in more protrusions and grooves. Long hydrophobic alkyl chains were self-assembled on the roughed surface during the mercaptan modification. A multilayer net-like surface, with protrusions and grooves together with a coral-like structure, was thereby obtained. Because most of the superhydrophobic surface area was covered by the air in the solid/liquid contact area, corrosive ions were prevented from reaching the magnesium surface, greatly improving the corrosion resistance. The contact-angle at the surface depended upon the various treatments (figures 15 to 17). The angle at the surface after polishing and cleaning was about 38°. It decreased to about 5° after etching, showing that the surface was rendered superhydrophilic by the etching. The surface remained superhydrophilic, with a contact-angle of about 8° after the nitrate treatment. The angles indicated superhydrophobicity following dodecyl mercaptan treatment. With increasing dodecyl mercaptan concentration, the contact-angle decreased after an initial increase: the angle attained 123° when the dodecyl mercaptan concentration was 0.01mol/*l*, but attained 153° when the dodecyl mercaptan concentration increased to 0.10mol/*l*. The angle was about 103° when the mercaptan concentration reached 0.5mol/l or more. The E_{corr} value of the alloy increased, and the I_{corr} value decreased with increasing water contact-angle (table 33), indicating that the instantaneous corrosion rate decreased with increasing contact-angle.

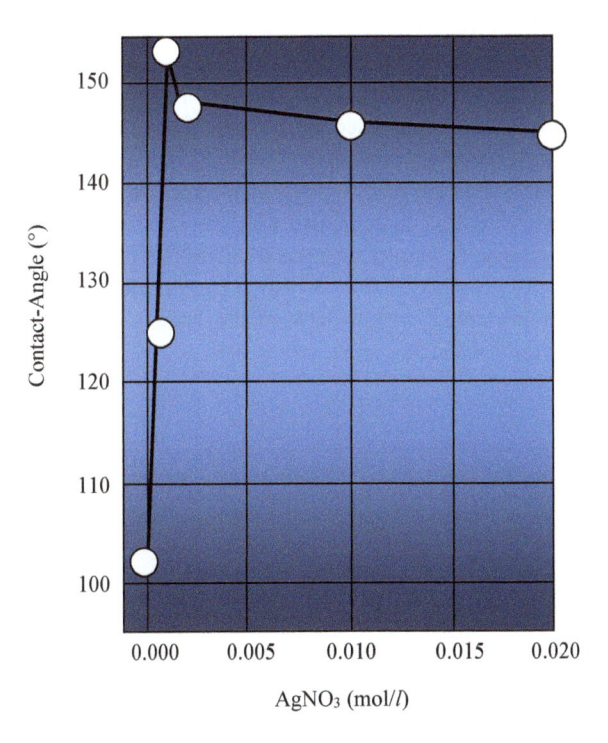

Figure 16. Effect of $AgNO_3$ concentration upon the water contact-angle of AZ91 surfaces

AZ91D

Stable superhydrophobic surfaces were created on AZ91D alloy by combining electrodeposition and chemical modification[125]. The as-prepared surface, having a pine-cone like hierarchical structure, had a water contact-angle of 163.3° and a sliding-angle of about 1.2°. The surface retained its superhydrophobicity following mechanical abrasion over 700mm. The as-prepared surface repelled corrosive and saline liquids. After exposure to the atmosphere for more than 240 days, the surfaces still had contact-angles of up to 158.5° and a sliding-angle of 6.0°. The surface provided excellent anti-corrosion protection to the alloy in neutral 3.5wt%NaCl, in that the corrosion rate was 0.003% of that of the bare alloy.

A high-temperature heating method was proposed for the preparation of a superhydrophobic layer with a coral-like surface structure on AZ91D[126]. The stearic acid modified superhydrophobic surface had a contact-angle of 159.1° and a rolling-angle of 4.8°. The corrosion current of the superhydrophobic surface decreased by some 2 orders-of-magnitude relative to that of the alloy substrate, and its inhibition efficiency was 96.94%. When the temperature was increased to 190°C the contact-angle remained above 150° (figure 18). When the heating temperature was 220C, the surface changed to the superhydrophobic state. Polarization tests were used to determine current density and corrosion potential data (table 34). As compared with untreated AZ91D, the corrosion potential of the superhydrophobic surface changed from -1.2525 to -1.1495$V_{Ag/AgCl}$, and the corrosion current density decreased from 3.1713×10^{-4} to $9.7053 \times 10^{-6} A/cm^2$. The superhydrophobic layer prevented contact between the NaCl solution and the alloy substrate, rendering corrosion difficult.

Table 34. Electrochemical data for AZ91D

Surface	E_{corr} ($V_{Ag/AgCl}$)	I_{corr} (A/cm^2)
normal	-1.2525	3.1713×10^{-4}
SHP	-1.1495	9.7053×10^{-6}

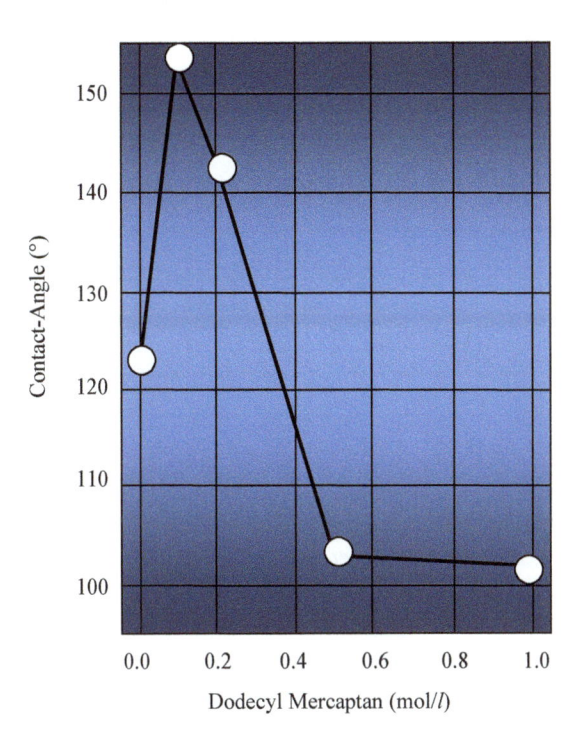

Figure 17. Effect of dodecyl mercaptan concentration upon the water contact-angle of AZ91 surfaces

MA8

Superhydrophobic coatings were produced on wrought MA8 by means of nanosecond laser processing, followed by the chemical vapour deposition of fluorosilane. The as-prepared coatings, exposed to 0.5M NaCl, exhibited a corrosion current which was more than 8 orders-of-magnitude lower than that for a polished sample which spent 2h in contact with the medium[127]. The initial value of the corrosion current density could be as low as 7×10^{-12} A/cm^2. Following 30 days of immersion, the current was 4 orders-of-magnitude lower. The surface, just after laser-texturing, was in a superhydrophilic state, due to its hierarchical roughness. The texture was made up of aggregates of magnesium oxide nano-particles which were deposited onto the surface from the laser plume during ablation. The as-formed magnesium oxide layer had a low density, yielding a Pilling-Bedworth ratio of 0.81 and consequently a non-protective porous layer which permitted the access of corrosive media to the magnesium/oxide interface. The water-droplet contact-angle and roll-off angle were determined after exposure to 0.5M NaCl (table 35).

Table 35. Wettability of superhydrophobic MA8 exposed to 0.5M NaCl

Surface	Contact-Angle (°)	Roll-Off Angle (°)
as-prepared	171.5	2.6
after 30 days of exposure	168.2	13.9

ZK60

A superhydrophobic surface was produced on pre-treated ZK60 by nano-silver deposition and electroless Ni-P plating, followed by stearic acid modification[128]. The surface had an hierarchical structure, and a water contact-angle of 157.8°. Immediately after the nano-silver deposition, the water contact-angle could attain 150.8°. Shortly after that, the contact-angle reached the 157.8° value. After a few more minutes, the contact-angle fell to 156.6°. At this point, many nano-silver were deposited which destroyed the micro-scale microstructure, and decreased the water contact-angle. When the reaction continued for 600s, the contact-angle decreased to 152.5°. Corrosion polarization data were determined for the alloy in 3.5wt%NaCl solution. The untreated alloy had the lowest corrosion potential and the maximum corrosion current density, but the worst corrosion resistance. Pre-treatments had little effect in improving the corrosion performance of the alloy following nano-silver deposition. Following electroless Ni-P plating, the prepared superhydrophobic material exhibited the highest corrosion potential and the smallest corrosion current density; indicating that it also promised the best corrosion resistance as the corrosion current density decreased by 2 orders-of-magnitude (table 36).

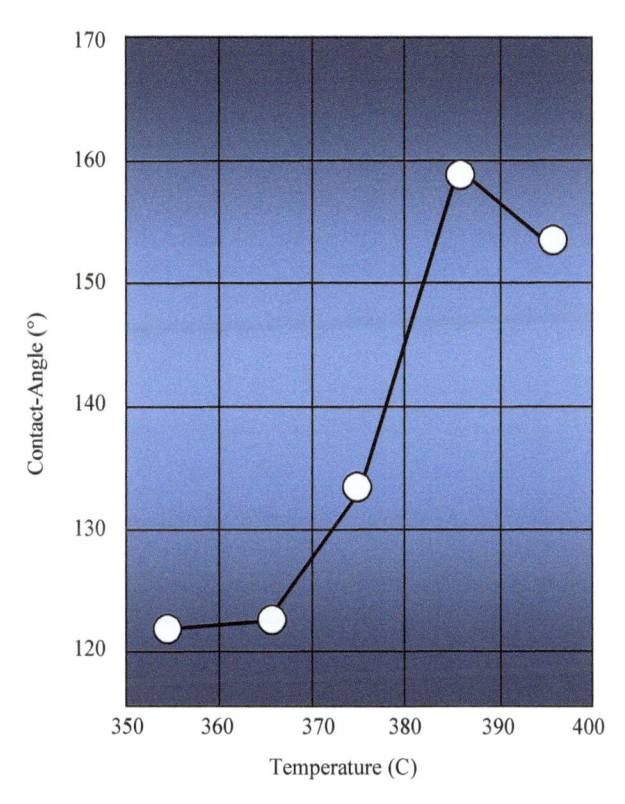

Figure 18. Contact-angle of AZ91D as a function of heating temperature

Table 36. Electrochemical parameters of ZK60 in 3.5wt%NaCl

Surface	E_{corr} (V)	I_{corr} (A/cm^2)
bare substrate	-1.544	4.960×10^{-5}
nano-silver deposited	-1.411	4.545×10^{-5}
pre-treated and nano-silver deposited	-1.326	2.746×10^{-5}
superhydrophobic	-0.117	3.051×10^{-7}

Nickel

Superhydrophobic nickel films were prepared on copper via electrodeposition in the presence of palmitic acid, plus egg-shell extract as an additive[129]. In the absence of the latter, the nickel film was smooth. In the presence of the egg-shell extract, the film had a micro-nano structure with a pine-cone shape. These surfaces had a contact-angle of 162° and sliding-angle of 3° (table 37). The superhydrophobicity of the deposited film increased when the deposition-current density was increased from 0.7 to 3.0mA/cm². This superhydrophobicity was retained after 200 abrasions. The films also exhibited a marked chemical stability in acidic and alkaline environments, and the as-prepared superhydrophobic film offered a corrosion protection of better than 99% (table 38).

Table 37. Contact-angles and sliding-angles on nickel surfaces

Current (mA/cm²)	Time (s)	Egg-shell	Contact-Angle (°)	Sliding-Angle (°)
0.7	1715	no	152	6.0
0.7	1715	yes	156	3.5
3.0	400	yes	162	3.0

Titanium

Superhydrophobic titanium surfaces were prepared by single-step anodization in an organic electrolyte, followed by stearic acid modification[130]. The material was anodized so as to produce micro-nano hierarchical morphology which was then treated with stearic acid. The morphology of the surfaces could be modified by varying the anodizing time from 1 to 5h and the voltage from 5 to 50V. The static water contact-angle and contact-angle hysteresis on the best superhydrophobic surface (anodized for 3h at 35V) were 160.1° and 7°, respectively.

Table 38. Corrosion data for nickel surfaces in 3.5wt%NaCl solution

Current (mA/cm^2)	Time (s)	Egg-shell	E$_{corr}$ (mV$_{SCE}$)	I$_{corr}$ (mA/cm^2)	P(%)
0.7	1715	no	225	0.00103	95.16
0.7	1715	yes	232	0.000621	97.08
3.0	400	yes	229	0.000208	99.02

Superamphiphobic surfaces could be prepared by means of 1-step anodization and then fluoroalkylsilane modification. The prepared alloy surfaces had water, glycerol and hexadecane contact-angles of 166.4°, 158.4° and 152.5°, respectively[131]. The corresponding sliding-angles were all within 10°. The observations were attributed to the re-entrant micro-nano structures and the low surface-energy modification. Exposure of the surfaces to ultra-violet light, immersion and abrasion showed that they possessed good stability under harsh conditions.

Superhydrophobic surfaces were prepared by anodization in sodium chloride solution, followed by immersion in perfluorodecyltriethoxysilane[132]. The 50nm anodic film comprised TiO_2 and $TiCl_3$ and had an hierarchical structure which consisted of a micro-scale horn structure and a nano-scale overlay. This surface had a water contact-angle of 151.9° and a sliding-angle of 3° following immersion. The surface coverage of the hierarchical structure was improved by mechanical attrition, which grain-refined the titanium. The thickness of 200nm, of the anodic film on the mechanically-treated surface, was clearly greater than the 50nm of the titanium surface. This was due to the large number of grain boundaries on the surface, which acted as rapid-diffusion paths during anodization. On the other hand, the adhesion of the mechanically-treated and anodized film was inferior to the film which was formed by anodization alone. This was attributed to a large number of pores within the former films.

A 2-step process was used to create superhydrophobic surfaces on N-purity titanium by shot-peening, followed by chemical etching[133]. The etched surface was then modified using methyltrichlorosilane in order to lower the surface energy. A nano-scale fibrous network was present on the surface, with chemical bonds existing between the functional groups of the methyltrichlorosilane and the titanium surface. The pre-shot peened surfaces had a maximum water contact-angle of 159° following methyltrichlorosilane modification.

Superhydrophobic surfaces with complex micro-pore structures and low surface roughness were created via anodic oxidation in NaOH-H_2O_2 solution. Fluoroalkylsilane was then used to reduce the surface energy of the electrochemically oxidized surface[134]. The as-prepared surfaces had a roughness of 0.669μm, with a water contact-angle of 158.5° and a tilting-angle of 5.3°, as well as offering good long-term stability and abrasion resistance.

Superhydrophobic surfaces were prepared on 4N-purity titanium by means of anodization and surface-energy modification[135]. The surface had rough micro-protrusions with a nano-flake morphology which resembled pine cones[136]. This microstructure, and the wettability of the surface, could be adjusted by altering anodization parameters such as the anodization time and voltage. The highest water contact-angle was 161.4° together with a sliding-angle which was essentially 0°. In further work it was noted that surfaces without chemical modification were superhydrophilic, but became superhydrophobic during exposure to air for long periods. Following high-temperature annealing, the time required for the superhydrophilic to superhydrophobic wettability transition, with a maximum water contact-angle of 153° and sliding-angle of essentially 0°, decreased from 81 days to 63 days. The wettability transition was due mainly to the adsorption of organic compounds from the ambient atmosphere and to air trapped in the microstructure. The superhydrophobic surfaces exhibited good anti-icing and self-cleaning behaviour and long-term stability.

Antireflective microstructures were created on superhydrophobic titanium alloy surfaces by means of nanosecond laser treatment[137]. By optimizing the laser speed, surfaces with a low reflectance over the visible range and near-infrared ranges (400 to 1000nm) were obtained. When the scanning-speed was 100mm/s, the minimum reflectance in the 900 to 1000nm wavelength range was less than 1%, and the total reflectance in the 500 to 1000nm range was below 6%. This was to be compared with the surface of bare titanium alloy, which absorbed about 50% of the incident light while the remaining 50% was reflected. The treated surfaces had a maximum contact-angle of 153.03°.

Pulsed nanosecond laser ablation was applied to titanium alloy, combined with functionalization using polysilazane, so as to obtain a biomimetic lotus-leaf superhydrophobic surface[138]. The as-synthesized surface had a water contact-angle of 164.1° and a sliding-angle of 1.5°. The addition of ZnO nano-particles to the organic coating markedly improved the corrosion resistance and anti-bacterial properties. The superhydrophobic surface offered an inhibition-rate of 93.89% with respect to *E. coli*.

Hierarchical binary surface structures were obtained by hydrothermal treatment with successive solutions of oxalic acid and sodium hydroxide[139]. The hierarchical surfaces, following fluoroalkylsilane modification, had a maximum contact-angle of 158.7° and a sliding-angle of 4.3°. This led to an efficient self-cleaning behaviour, with bouncing and rolling-off of water droplets from the surface. Exposure of the coating to 3.5wt%NaCl solution showed that it could retain its superhydrophobic state for 48h and thus prevent ingress of the corrosive medium.

Table 39. Contact-angles on titanium for various $AgNO_3$ immersion times

Time (h)	Contact-Angle (°)
0.5	144.7
2	151
5	152.7
7	154
12	152.3

Superhydrophobic surfaces were prepared on 2N8-purity titanium which was pre-treated by mechanical polishing and anodizing, or by mechanical polishing alone. This was combined with the self-assembly of polydopamine and silver nano-particles, and post-modification using 1H,1H,2H,2H-perfluorodecanethiol[140]. The anodizing process could in fact be eliminated. The hydrophobicity increased with increasing deposition-time in silver nitrate solution (table 39). Surfaces which were so treated for 7h exhibited the optimum hydrophobic effect, with a water contact-angle of up to 154°. The surface was quite rough and was covered by relatively uniform micro-nano silver structures. The good hydrophobicity was attributed to the rough hierarchical microstructure, together with a low surface energy. The polarization curves of samples with and without a superhydrophobic surface were similar. Both had passivation characteristics, indicating that a stable passive film was formed on the surface. The surface with the superhydrophobic film had more positive E_{corr} and lower I_p as compared with that of the bare material (table 40).

Table 40. Corrosion potential passive current density of titanium

Surface	E_{corr} (V_{SCE})	I_p (nA/cm^2)
bare	-0.135	8.02
superhydrophobic	-0.0126	1.22

A micro-nano scale quasi-periodic self-organized structure was produced on titanium surfaces, using femtosecond laser ablation, which mimicked the surface of *Nelumbo nucifera* (lotus) leaves[141]. One scale consisted of large grain-like convex features which were between 10 and 20μm in size. The other feature, on the surface of the grains, comprised 200nm-wide irregular undulations. The use of these biomimetic surface patterns markedly changed the wettability of the surface. The original surface had a water contact-angle of 73°, but the laser-treated surface became superhydrophobic with a contact-angle of 166°. The interaction of *S. aureus* and *P. aeruginosa* with the superhydrophobic surfaces offered highly-selective retention of those pathogenic bacteria. The *S. aureus* cells could colonize the superhydrophobic surfaces, but *P. aeruginosa* cells were unable to attach to the surface.

Superhydrophobic surfaces were produced by using an ethanolic solution of myristic acid and hydrochloric acid plus simultaneous anodization and adsorption[142]. Using the optimum anodization potential, the surface was densely populated with hierarchical micro-nano clusters of titanium dioxide with adsorbed myristic acid. The maximum water contact-angle was 176.3°, with a sliding-angle of 1°. The surfaces had layered ridges which were decorated with sub-micron aggregates. The liquid/air areal fraction on the prepared superhydrophobic surface was about 0.97, with an asperity-slope greater than 71°. Superhydrophobic surfaces which were exposed to microbial cultures for 48h exhibited a 50% reduction in bacterial adhesion. Under a 15V potential, the modified surface had a slightly increased contact-angle of 110.8°, as compared to the 67.6° of the bare specimen. When the potential was increased to 20V, the surface was now superhydrophobic, a contact-angle of 151.3°. Specimens which were anodized at 30V and 40V had contact-angles of 160.6° and 169.2°, respectively. The maximum angle was 174.4°.

Superhydrophobic surfaces were produced on 3N-purity material by using rapid-breakdown anodization, combined with stearic acid[143]. The anodized surfaces exhibited islands of TiO_2 micro-clusters, with a complex hierarchical structure, which were randomly distributed in a passive TiO_2 matrix[144]. Surfaces which were anodized at 30V were hydrophobic, with a water contact-angle of about 130° and a sliding-angle greater than 30°. Surfaces which were anodized at 50V were superhydrophobic, with a contact-angle of about 154° and a sliding-angle of 10° after stearic acid modification. Sustained oxidation, and pore-growth via chemical or field-assisted dissolution reactions, controlled the final morphology of anodized surfaces. The superhydrophobic surface exhibited excellent self-cleaning properties and abrasion-resistance. When using molten stearic acid rather than a stearic acid-ethanol solution, the maximum water contact-angle was 167.8°, with a sliding-angle of 6°, on surfaces which had been anodized at 50V for 10min following molten stearic acid immersion at 125C for 0.5h. The same TiO_2 micro-clusters, with complex hierarchical structures and nano-pores were observed, and the micro-clusters were shown to be polycrystalline. Prolonged anodization promoted micro-cluster growth in each of their dimensions. The surfaces also exhibited vertical micro-plate crystals of stearic acid on the micro-clusters, and this increased the contact-angle. The stearic acid bonded with the micro-clusters via surface hydroxyl groups.

Stable superhydrophobic surfaces, offering good stretch-resistance, were produced by chemical etching[145]. The best examples had a water contact-angle of 164° together with a water tilting-angle of about 2°, and also possessed a marked stretch-resistance.

Ti-6Al-4V

Superhydrophobic surfaces were prepared on Ti-6Al-4V by means of high-speed micro-milling, anodic oxidation and fluoroalkylsilane modification[146]. Regular microgrooves were constructed by micro-milling and nano-tube arrays were created via anodic oxidation. Fluoroalkylsilane was then used to self-assemble a monolayer on the surface which had a micro-nano structure. Unlike the polished surfaces, the modified samples were superhydrophobic, with a water contact-angle of 153.7° and a contact-angle hysteresis of 2.1° (advancing and receding contact-angles of 153.8° and 151.7°, respectively). The surface of the polished material was smooth, while that of the machined material exhibited high and low staggered peaks and valleys, with traces of the milling at the bottom of the grooves and some processing burrs on the surface. Following anodic oxidation, closely-packed nano-tubes were present on the surface of the machined material, having an inner diameter of 25 to 30nm. They did not affect the structures produced by the micro-milling. Following fluorination, the surface morphology was not greatly changed.

A 1064nm pulsed picosecond-laser was used to create a micro-nano hierarchical structure on Ti-6Al-4V. The initial contact-angle of the polished surface was 71.6°. The imposed structure comprised dimple arrays having various diameters, depths and areal densities which were produced by controlling the pulse-energy and the number of pulses[147]. The contact-angle of the laser-textured surface was less than 30° and all of the surfaces were highly hydrophilic. The pitch between micro-dimples was 100, 80 or 60μm, corresponding to areal densities of 13, 20 or 35%, respectively. The aspect-ratio was 50%. The nano-features consisted of so-called laser-induced periodic surface structures, the dimensions of which could be varied by changing the laser energy-density and scanning-speed. The ripples had a period of about 1100nm when the energy-densities and scanning-speeds were 0.107 to $0.218J/cm^2$ and 30 to 50mm/s. The contact-angle increased as the density of micro-textured surfaces increased. There was a slow (up to 4 weeks) transition from hydrophilic to hydrophobic. The wetting of the textured surfaces could be described in terms of the Cassie model. Following low-temperature annealing, the slow surface wettability transition could be markedly accelerated. This was attributed to changes in hydroxyl groups on the surface. The contact-angle of smooth surfaces did not change following the annealing treatment, indicating that it had little effect upon the wettability of the smooth surface. The contact-angle of textured surfaces was markedly changed. The contact-angle of surfaces with micro-dimples increased from 78.49° to 107.52°. The contact-angle of the rippled surfaces increased from 87.08° to 106.53° The angle increased from 101.27° to 136.79° when surfaces with micro-dimples were further covered with ripples. When the dimple surfaces were covered with periodic ripples, superhydrophobic surfaces with a contact-angle of up to 144.58° could be obtained.

Nanostructured superhydrophobic surfaces on Ti-6Al-4V were prepared by means of laser and anodizing treatments. The laser treatment generated a rough surface with parallel grooves and protrusions, offering superhydrophobicity following organic modification[148]. The anodizing treatment created a titanium dioxide nano-tube film. As compared with samples which were only laser-treated, the oxide nano-tube film improved the corrosion resistance (table 41) and mechanical stability of the superhydrophobicity. The water contact-angle of the untreated alloy surface was 72°; meaning a hydrophilic surface. Following the laser treatment, the angle decreased to 0°; meaning a superhydrophilic surface. Following the anodizing treatment, the angle remained equal to 0°. The Wenzel model explained the above phenomena, in that the hydrophilicity of the hydrophilic surface increased with increasing surface roughness. The laser treatment produced a rough alloy surface, thus greatly increasing the surface roughness and reducing the contact-angle. The use of fluorosilane modification reduced the surface free-energy provided superhydrophobicity. Following modification, the contact-angle was 103° for untreated material, thus transforming the original hydrophilic surface into a hydrophobic surface. Laser-treatment produced superhydrophobic surfaces with a contact-angle of 154°, and anodizing further increased the angle to 158°. The chemical stability of the superhydrophobicity was tested by immersing the material in 3.5wt%NaCl solution for 12h. The contact-angle of the superhydrophobic surfaces hardly decreased after 12h, and this was attributed to the high water-repellency of the surface. A linear abrasion test was used to determine the mechanical stability of the superhydrophobic surfaces by rubbing them over 600-grit silicon carbide paper under a load of 100g load at a constant speed. After 5 cycles, the laser-treated superhydrophobic surfaces retained a contact-angle of more than 150°. The angle decreased to 141° when the test-distance reached 2000mm after 10 cycles, but the surface continued to retain good hydrophobicity. The laser treatment turned the morphology into one consisting of peaks and valleys. The uneven structure could accommodate wear debris, reduce abrasive wear and improve the mechanical stability of surface features. A microstructure with evenly distributed peaks also offered a better mechanical performance and maintained superhydrophobicity. The anodizing further improved the mechanical stability of the superhydrophobicity. The surface retained superhydrophobicity after 6 testing cycles, and the contact-angle was 147° when the testing distance reached 2000mm. The improved mechanical behaviour was attributed to the oxide nano-tube layer on the laser-treated surface. This layer could better accommodate debris during wear tests. The nano-tube structure itself was difficult to remove. The unworn nano-tube structure could also offer an extra air-cushion effect for aiding superhydrophobicity.

Table 41. Corrosion parameters of Ti-6Al-4V after various treatments

Treatment	E_{corr} (V_{SCE})	I_{corr} (A/cm^2)	Corrosion-Rate (mm/y)
none	-0.243	1.37 x 10^{-5}	0.161
laser	-0.289	9.88 x 10^{-6}	0.116
laser+anodizing	-0.184	4.72 x 10^{-6}	0.055
laser+fluorosilane	-0.168	4.34 x 10^{-8}	0.0005096
laser+anodizing+fluorosilane	-0.01	7.66 x 10^{-9}	0.00008994

Superhydrophobic lead surfaces were produced on Ti-6Al-4V substrates by 30 to 60s of immersion, followed by modification using low surface-energy materials. Hierarchical micro-nano scale structures were generated by the immersion in Pb(CH$_3$COO)$_2$ solution, and were then modified using fluoroalkylsilane in order to reduce the surface energy[149]. The coatings had a water contact-angle of 165.5° and a sliding-angle of 4.6°, and exhibited long-term stability in air together with good self-cleaning and anti-corrosion properties.

Films of titanium dioxide with various nanostructures were produced on Ti-6Al-4V and commercial-purity titanium by means of electrochemical anodization[150]. The oxide films prepared on commercial-purity surfaces had a porous structures, while those on the alloy consisted of nano-rods. The oxide nano-film on the commercial material were hydrophilic, with a contact-angle of about 19°. The films on the alloy were superhydrophilic, with a contact-angle of less than 2°. Following treatment with fluorinated silane, the self-assembled films were now superhydrophobic, with contact-angles of 150° and 158°, respectively. The nanostructures and fluoroalkysilane were both thought to play important roles in controlling the wettability. Surfaces having nano-roughness alone did not exhibit superhydrophobicity.

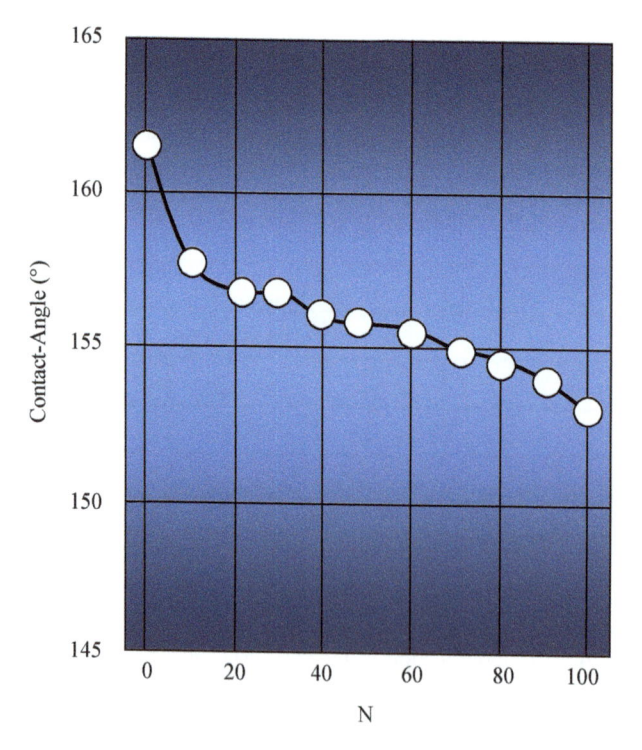

Figure 19. Contact-angle of tungsten surface as a function of the number of abrasion cycles

A study of the wetting state of liquid droplets on a Ti-6Al-4V micro-nano hierarchical hydrophobic surface clarified the transition from Wenzel to Cassie behaviour. Theory and experiment showed that a 1-dimensional nano-wire structure, which was produced on the surface by a hydrothermal treatment, changed the wetting-state of liquid droplets from Wenzel to Cassie due to its good dimensional correspondence to the micro-scale structure[151]. This increased the apparent contact-angle of liquid droplets on the solid surface to 161°, and also greatly decreased the sliding-angle, to 3° and the contact-angle hysteresis to 2°.

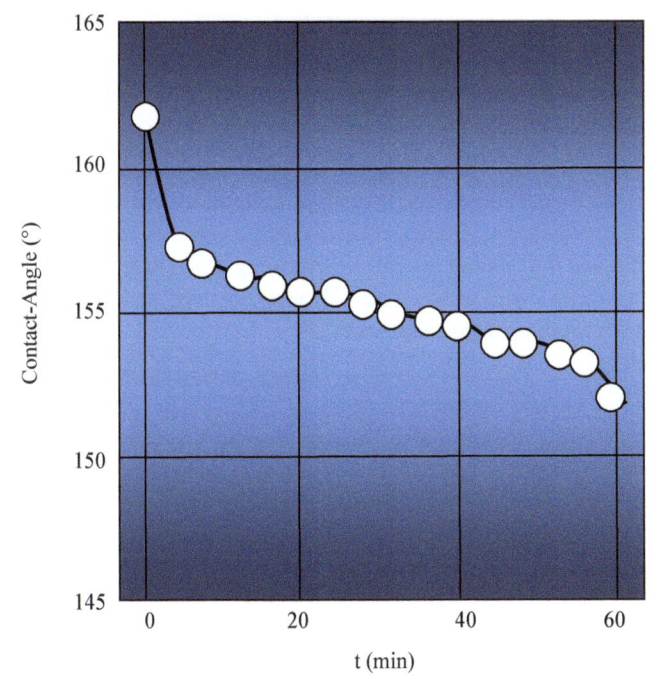

Figure 20. Contact-angle of tungsten surface as a function of particle impact time

Anti-icing superhydrophobic surfaces were produced, on Ti-6Al-4V, which possessed an hierarchical structure comprising micro-scale arrays that were created by micro-machining and nano-hairs that were created by hydrothermal growth[152]. The superhydrophobic surfaces exhibited high non-wettability, with a contact-angle of up to

160° and a contact-angle hysteresis of 2°. This led to an icing delay-time of some 765s, thus hindering ice-formation and growth at -10C and limiting the ice adhesion-strength to 70kPa.

Tungsten

One-step nanosecond laser scanning was used to produce microstructures on tungsten surfaces. A higher laser fluence, together with a smaller scanning interval and a lower scanning speed were best for obtaining lotus-leaf like hierarchical microstructures[153]. When the fluence was greater than $1.8J/cm^2$, for a scanning speed of 10.0mm/s and an interval of 1.0μm, the leaf-like hierarchical microstructures together with fluoroalkylsilane modification led to a contact-angle, for 5μ*l* water droplets, of 162° and a minimum rolling angle of 1.0°.

Superhydrophobic coatings on the surface of tungsten were prepared by nanosecond laser treatment and the deposition of fluoro-oxysilane from the vapour. Varying the time used to pre-treat the surface with oxygen plasma made it possible to adjust the density of adsorption sites and thus control the chemical stability of the hydrophobicity modifier[154]. Surfaces with contact-angles greater than 170° were obtained, and these were maintained during long-term exposure to aqueous media.

An ultra-fast laser was used to create micro-nano hierarchical structures which featured micro-cones[155]. The heights and separations of such features were important for the resistance to tangential abrasion and dynamic impact, respectively (figures 19 to 21). Superhydrophobic tungsten hierarchical surfaces were prepared which could withstand 70 abrasion cycles, 28min of solid-particle impact or 500 tape-peeling cycles, while maintaining a contact-angle greater than 150° and sliding-angles of less than 20°. In the linear abrasion tests, the tested surfaces were rubbed against 1000-grit sandpaper under a load of 1.2kPa with each cycle involving a to-and-fro relative motion of 10cm. In the solid-particle impact tests, the superhydrophobic surfaces were placed on a substrate which was tilted at 45° while silica-sand with a particle-size of 100 to 300μmm was released, from a height of 25cm, onto the surface at a flow-rate of 10g/min. In the tape-peeling tests, tape with an adhesive strength of 710N/m was used.

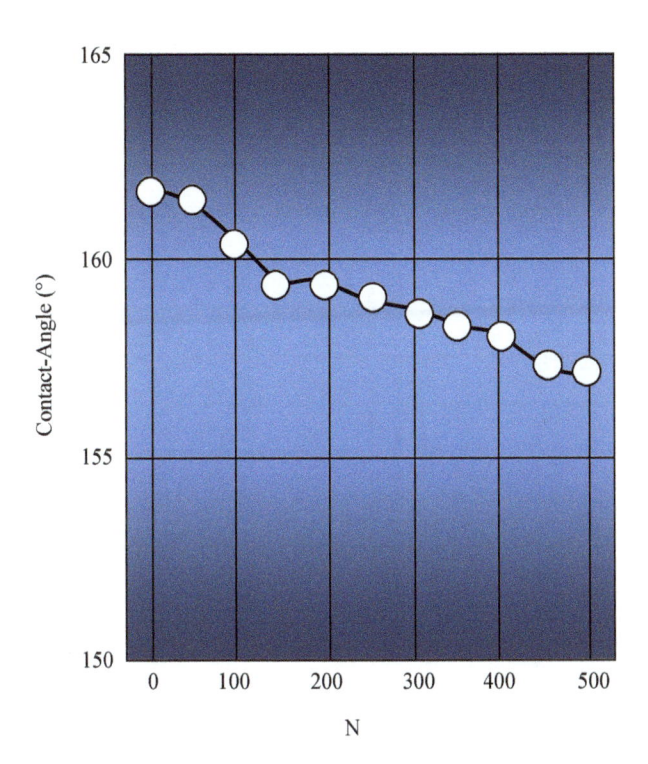

Figure 21. Contact-angle of tungsten surface as a function of the number of tape-peeling cycles

In later studies, superhydrophobic surfaces with hierarchical roughness were again created by pulsed nanosecond infra-red (1064nm) laser treatment, followed by the chemical vapour deposition of fluoro-oxysilane[156]. The pulse-duration ranged from 4 to 200ns and the pulse-frequency from 20 to 100kHz, while the energy of a single pulse could be up to 0.95mJ. The beam spot meanwhile was about 40μm in diameter and was moved in 2 mutually perpendicular directions. The durability of the coated surfaces depended upon the surface morphology and composition; qualities which could be controlled by adjusting the laser-treatment parameters. The laser-texturing of the surface led to a decrease in the contact-angle from 58°, which was expected for flat bare metal to complete wetting, and an angle of 0°. Following the deposition of fluoro-oxysilane, the surfaces became highly hydrophobic (table 42). Quite wide variations in the laser-treatment parameters only slightly affected the resultant contact-angle. Use of the optimum treatment parameters led to a contact-angle of 172.1° and a roll-off angle of 1.5°. These superhydrophobic properties were retained after oscillating sand-abrasion for 10h, contact with water droplets for more than 50h and after several cycles of falling-sand tests

Table 42. Contact-angles on tungsten as a function of laser parameters

v_s (mm/s)	d_s (/mm)	N	F (J/cm^2)	CA (°)	RO (°)
500	12.5	1	71	142.1	-
250	12.5	1	143	147.7	-
100	12.5	1	356	170.5	1.6
100	40	1	1140	170.8	2.2
100	75	1	2138	171.4	2.1
500	25	1	143	170.2	7.8
500	25	2	285	171.0	4.5
500	25	5	715	172.2	3.5
250	25	1	285	170.3	7.9

v_s: scan velocity, d_s: scan line-density, N: number of passes, F: fluence

About the Author

Dr. Fisher has wide knowledge and experience of the fields of engineering, metallurgy and solid-state physics, beginning with work at Rolls-Royce Aero Engines on turbine-blade research, related to the Concord supersonic passenger-aircraft project, which led to a BSc degree (1971) from the University of Wales. This was followed by theoretical and experimental work on the directional solidification of eutectic alloys having the ultimate aim of developing composite turbine blades. This work led to a doctoral degree (1978) from the Swiss Federal Institute of Technology (Lausanne). He then acted for many years as an editor of various academic journals, in particular *Defect and Diffusion Forum*. In recent years he has specialized in writing monographs which introduce readers to the most rapidly developing ideas in the fields of engineering, metallurgy and solid-state physics. He is co-author of the widely-cited student textbook, *Fundamentals of Solidification*. Google Scholar credits him with **8687** citations and a lifetime h-index of 14.

References

[1] Snoeijer J.H., Brunet P., American Journal of Physics, 80, 2012, 764-771. https://doi.org/10.1119/1.4726201

[2] Watson G.S., Gellender M., Watson J.A., Biofouling, 30[4] 2014, 427-434. https://doi.org/10.1080/08927014.2014.880885

[3] Cao M., Li K., Dong Z., Yu C., Yang S., Song C., Liu K., Jiang L., Advanced Functional Materials. 25, 2015, 4114–4119. https://doi.org/10.1002/adfm.201501320

[4] Huang S., Song J., Lu Y., Lv C., Zheng H., Liu X., Jin Z., Zhao D., Carmalt C.J., Parkin I.P., Journal of Materials Chemistry A, 4, 2016, 13771-13777. https://doi.org/10.1039/C6TA04908G

[5] Zhang B., Zhu Q., Li Y., Hou B., Chemical Engineering Journal, 352, 2018, 625-633. https://doi.org/10.1016/j.cej.2018.07.074

[6] Zhang B., Xu W., Zhu Q., Li Y., Hou B., Journal of Colloid and Interface Science, 532, 2018, 201-209. https://doi.org/10.1016/j.jcis.2018.07.136

[7] Nakajima D., Kikuchi T., Natsui S., Suzuki R.O., Applied Surface Science, 440, 2018, 506-513. https://doi.org/10.1016/j.apsusc.2018.01.182

[8] Escobar A.M., Llorca-Isern N., Applied Surface Science, 305, 2014, 774-782. https://doi.org/10.1016/j.apsusc.2014.03.196

[9] Zheng S., Li C., Fu Q., Hu W., Xiang T., Wang Q., Du M., Liu X., Chen Z., Materials and Design, 93, 2016, 261-270. https://doi.org/10.1016/j.matdes.2015.12.155

[10] Abdolmaleki M., Allahgholipour G.R., Tahzibi H., Azizian S., Materials Chemistry and Physics, 313, 2024, 128711. https://doi.org/10.1016/j.matchemphys.2023.128711

[11] Yin L., Wang Y., Ding J., Wang Q., Chen Q., Applied Surface Science, 258[8] 2012, 4063-4068. https://doi.org/10.1016/j.apsusc.2011.12.100

[12] Liu L., Zhao J., Zhang Y., Zhao F., Zhang Y., Journal of Colloid and Interface Science, 358[1] 2011, 277-283. https://doi.org/10.1016/j.jcis.2011.02.036

[13] Chen Z., Guo, Y., Fang S., Surface and Interface Analysis, 42[1] 2010, 1-6. https://doi.org/10.1002/sia.3126

[14] Jagdheesh R., García-Ballesteros J.J., Ocaña J.L., Applied Surface Science, 374, 2016, 2-11. https://doi.org/10.1016/j.apsusc.2015.06.104

[15] Liu C., Su F., Liang J., RSC Advances, 4, 2014, 55556-55564. https://doi.org/10.1039/C4RA09390A

[16] Forooshani H.M., Aliofkhazraei M., Rouhaghdam A.S., Journal of the Taiwan Institute of Chemical Engineers, 72, 2017, 220-235. https://doi.org/10.1016/j.jtice.2017.01.014

[17] Song Y., Wang C., Dong X., Yin K., Zhang F., Xie Z., Chu D., Duan J., Optics and Laser Technology, 102, 2018, 25-31. https://doi.org/10.1016/j.optlastec.2017.12.024

[18] Rasitha T.P., Vanithakumari S.C., Krishna D.N.G., George R.P., Srinivasan R., Philip J., Progress in Organic Coatings, 162, 2022, 106560. https://doi.org/10.1016/j.porgcoat.2021.106560

[19] Tong W., Cui L., Qiu R., Yan C., Liu Y., Wang N., Xiong D., Journal of Materials Science and Technology, 89, 2021, 59-67. https://doi.org/10.1016/j.jmst.2021.01.084

[20] Song M., Liu Y., Cui S., Liu L., Yang M., Applied Surface Science, 283, 2013, 19-24. https://doi.org/10.1016/j.apsusc.2013.05.088

[21] Lu S., Chen Y., Xu W., Liu W., Applied Surface Science, 256[20] 2010, 6072-6075. https://doi.org/10.1016/j.apsusc.2010.03.122

[22] Jafari R., Farzaneh M., Materials Science Forum, 706-709, 2012, 2874-2879. https://doi.org/10.4028/www.scientific.net/MSF.706-709.2874

[23] Lu Y., Shen Y., Tao J., Wu Z., Chen H., Jia Z., Xu Y., Xie X., Langmuir, 36, 2020, 880-888. https://doi.org/10.1021/acs.langmuir.9b03411

[24] Sheng X., Zhang J., Applied Surface Science, 257[15] 2011, 6811-6816. https://doi.org/10.1016/j.apsusc.2011.03.002

[25] Wang Y., Shi H., Li X., Journal of Materials Science, 54, 2019, 7469-7482. https://doi.org/10.1007/s10853-018-03273-y

[26] Parin R., Del Col D., Bortolin S., Martucci A., Journal of Physics - Conference Series 745, 2016, 032134. https://doi.org/10.1088/1742-6596/745/3/032134

[27] Zhong Z.W., Niu J.L., Ma W., Yao S.H., Yang M., Wang Z.K., Journal of Physics - Conference Series, 2069, 2021, 012121. https://doi.org/10.1088/1742-6596/2069/1/012121

[28] Lv S., Zhang X., Yang X., Liu X., Yang Z., Zhai Y., Materials Research Express, 9, 2022, 026520. https://doi.org/10.1088/2053-1591/ac433a

[29] Du X.Q., Chen Y., Materials Research Express, 7, 2020, 056405. https://doi.org/10.1088/2053-1591/ab9253

[30] Wang H., Dai D., Wu X., Applied Surface Science, 254[17] 2008, 5599-5601. https://doi.org/10.1016/j.apsusc.2008.03.004

[31] Rezayi T., Entezari M.H., Journal of Colloid and Interface Science, 463, 2016, 37-45. https://doi.org/10.1016/j.jcis.2015.10.029

[32] Saffari H., Sohrabi B., Noori M.R., Bahrami H.R.T., Applied Surface Science, 435, 2018, 1322-1328. https://doi.org/10.1016/j.apsusc.2017.11.188

[33] Zuo Z., Liao R., Guo C., Yuan Y., Zhao X., Zhuang A., Zhang Y., Applied Surface Science, 331, 2015, 132-139. https://doi.org/10.1016/j.apsusc.2015.01.066

[34] Xu S., Wang Q., Wang N., Colloids and Surfaces A, 595, 2020, 124719. https://doi.org/10.1016/j.colsurfa.2020.124719

[35] Volpe, A., Covella, S., Gaudiuso C., Ancona A., Coatings, 11[3] 2021, 369. https://doi.org/10.3390/coatings11030369

36 Yang H., Gao Y., Qin W., Sun J., Huang Z., Li Y., Li B., Sun J., Journal of Alloys and Compounds, 898, 2022, 163038. https://doi.org/10.1016/j.jallcom.2021.163038

[37] Fan C., Wang X., Liu Y., Li C., Liu X., Journal of Physics - Conference Series, 2174, 2022, 012050. https://doi.org/10.1088/1742-6596/2174/1/012050

[38] Da Silva R.G.C., Vieira M.R.S., Malta M.I.C., Da Silva C.H., De Oliveira S.H., Filho S.L.U., Surface and Coatings Technology, 369, 2019, 311-322. https://doi.org/10.1016/j.surfcoat.2019.04.040

[39] Zhao Q., Tang T., Wang F., Coatings, 8, 2018, 390. https://doi.org/10.3390/coatings8110390

[40] Calabrese L., Khaskhoussi A., Patane S., Proverbio E., Coatings, 9, 2019, 352. https://doi.org/10.3390/coatings9060352

[41] Fahim J., Ghayour H., Hadavi S.M.M., Tabrizi S.A.H., Protection of Metals and Physical Chemistry of Surfaces, 54[5] 2018, 899-908. https://doi.org/10.1134/S2070205118050052

[42] Song X.G., Liang Z.H., Wang H.J., Hu S.P., Fu W., Xu X.R., Tan C.W., Journal of Coating Technology Research, 20[6] 2023, 1897-1912. https://doi.org/10.1007/s11998-023-00785-4

[43] Dong X., Meng J., Hu Y., Wei X., Luan X., Zhou H., Micromachines, 11, 2020, 159. https://doi.org/10.3390/mi11020159

[44] Yang X., Liu X., Li J., Huang S., Song J., Xu W., Micro and Nano Letters, 10[7] 2015, 343-346. https://doi.org/10.1049/mnl.2014.0635

[45] Liu E., Zhu G., Dai P., Liu L., Yu S., Wang B., Xiong W., Colloids and Surfaces A, 652, 2022, 129916. https://doi.org/10.1016/j.colsurfa.2022.129916

[46] Lv F.Y., Zhang P., Applied Surface Science, 321, 2014, 166-172. https://doi.org/10.1016/j.apsusc.2014.09.147

[47] Barthwal S., Lee B., Lim S.H., Applied Surface Science, 496, 2019, 143677. https://doi.org/10.1016/j.apsusc.2019.143677

[48] Mora J., García P., Agüero A., Borrás A., González-Elipe A.R., López-Santos C., Applied Materials Today, 21, 2020, 100815. https://doi.org/10.1016/j.apmt.2020.100815

[49] Li X., Zhang Q., Guo Z., Shi T., Yu J., Tang M., Huang X., Applied Surface Science, 342, 2015, 76-83. https://doi.org/10.1016/j.apsusc.2015.03.040

[50] Zhang X., Zhao J., Mo J., Sun R., Li Z., Guo Z., Colloids and Surfaces A, 567, 2019, 205-212. https://doi.org/10.1016/j.colsurfa.2019.01.046

[51] Sun R., Zhao J., Li Z., Mo, J., Pan Y., Luo D., Progress in Organic Coatings, 133, 2019, 77-84. https://doi.org/10.1016/j.porgcoat.2019.04.020

[52] Feng L., Yan Z., Shi X., Sultonzoda F., Applied Physics A, 124, 2018, 142. https://doi.org/10.1007/s00339-017-1509-x

[53] Khaskhoussi A., Calabrese L., Proverbio E., Applied Science, 10, 2020, 2656. https://doi.org/10.3390/app10082656

[54] Liu Y., Li X., Yan Y., Han Z., Ren L., Surface and Coatings Technology, 331, 2017, 7-14. https://doi.org/10.1016/j.surfcoat.2017.10.032

[55] Boinovich L.B., Emelyanenko A.M., Modestov A.D., Domantovsky A.G., Shiryaev A.A., Emelyanenko K.A., Dvoretskaya O.V., Ganne A.A., Corrosion Science, 112, 2016, 517-52. https://doi.org/10.1016/j.corsci.2016.08.019

[56] Peng H., Luo Z., Li L., Xia Z., Du J., Zheng B., Materials Research Express, 6[9] 2019, 096586. https://doi.org/10.1088/2053-1591/ab3173

[57] Su F., Yao K., Liu C., Huang P., Journal of the Electrochemical Society, 160[11] 2013, D593. https://doi.org/10.1149/2.047311jes

[58] Alinezhadfar M., Abad S.N.K., Mozammel M., Surfaces and Interfaces, 21, 2020, 100744. https://doi.org/10.1016/j.surfin.2020.100744

[59] Daneshnia A., Raeissi K., Salehikahrizsangi P., Journal of Alloys and Compounds, 948, 2023, 169767. https://doi.org/10.1016/j.jallcom.2023.169767

[60] Liu W., Xu Q., Han J., Chen X., Min Y., Corrosion Science, 110, 2016, 105-113. https://doi.org/10.1016/j.corsci.2016.04.015

[61] Huang Y., Sarkar D.K., Chen X.G., Materials Letters, 64[24] 2010, 2722-2724. https://doi.org/10.1016/j.matlet.2010.09.010

[62] Guo Z., Fang J., Wang L., Liu W., Thin Solid Films, 515[18] 2007, 7190-7194. https://doi.org/10.1016/j.tsf.2007.02.100

[63] Wan Y., Chen M., Liu W., Shen X., Min Y., Xu Q., Electrochimica Acta, 270, 2018, 310-318. https://doi.org/10.1016/j.electacta.2018.03.060

[64] Shu Y., Lu X., Liang Y., Su W., Gao W., Yao J., Niu Z., Lin Y., Xie Y., Surface and Coatings Technology, 441, 2022, 128514. https://doi.org/10.1016/j.surfcoat.2022.128514

[65] Feng L., Yang M., Shi X., Liu Y., Wang Y., Qiang X., Colloids and Surfaces A, 508, 2016, 39-47. https://doi.org/10.1016/j.colsurfa.2016.08.017

[66] Kuang Y., Jiang F., Zhu T., Wu H., Yang X., Li S., Hu C., Materials Letters, 303, 2021, 130579. https://doi.org/10.1016/j.matlet.2021.130579

[67] Bahrami H.R.T., Ahmadi B., Saffari H., Materials Letters, 189, 2017, 62-65. https://doi.org/10.1016/j.matlet.2016.11.076

[68] Kumari P., Kumar A., Materials Today Communications, 36, 2023, 106744. https://doi.org/10.1016/j.mtcomm.2023.106744

[69] Yuan Z., Bin J., Wang X., Peng C., Wang M., Xing S., Xiao J., Zeng J., Xiao X., Fu X., Chen H., Surface and Coatings Technology, 254, 2014, 151-156. https://doi.org/10.1016/j.surfcoat.2014.06.004

[70] Bahrami H.R.T., Ahmadi B., Saffari H., Materials Research Express, 4[5] 2017, 055014. https://doi.org/10.1088/2053-1591/aa6c3b

[71] Chen J., Guo J., Qiu M., Yang J., Huang D., Wang X., Ding Y., Materials Transactions, 59, 2018, 5.

[72] Chaitanya B., Gunjan M.R., Sarangi R., Raj R., Thakur A.D., Materials Chemistry and Physics, 278, 2022, 125667. https://doi.org/10.1016/j.matchemphys.2021.125667

[73] Long J., Fan P., Zhong M., Zhang H., Xie Y., Lin C., Applied Surface Science, 311, 2014, 461-467. https://doi.org/10.1016/j.apsusc.2014.05.090

[74] Shu Y., Lu X., Lu W., Su W., Wu Y., Wei H., Xu D., Liang J., Xie Y., Surface and Coatings Technology, 455, 2023, 129216. https://doi.org/10.1016/j.surfcoat.2022.129216

[75] Shi X., Zhao L., Wang J., Feng L., Journal of Nanoscience and Nanotechnology, 20[10] 2020, 6317-6325. https://doi.org/10.1166/jnn.2020.17891

[76] Zhao Y., Zhang H., Wang W., Yang C., International Journal of Heat and Mass Transfer, 127[C] 2018, 280-288. https://doi.org/10.1016/j.ijheatmasstransfer.2018.07.153

[77] Song J., Xu W., Lu Y., Fan X., Applied Surface Science, 257 [24] 2011, 10910-10916. https://doi.org/10.1016/j.apsusc.2011.07.140

[78] Lv Y., Liu M., Surface Engineering, 35[6] 2019, 542-549. https://doi.org/10.1080/02670844.2018.1433774

[79] Jia C., Zhu J., Zhang L., Coatings, 12, 2022, 442. https://doi.org/10.3390/coatings12040442

[80] Vanithakumari S.C., George R.P., Mudali U.K., Philip J., Transactions of the Indian Institute of Metals, 72[5] 2019, 1133-1143. https://doi.org/10.1007/s12666-019-01586-3

[81] Hassan L.B., Saadi N.S., Karabacak T., International Journal of Advanced Manufacturing Technology, 93, 2017, 1107-1114. https://doi.org/10.1007/s00170-017-0584-7

[82] Yao C.W, Sebastian D., Lian I., Günaydın-Sen O., Clarke R., Clayton K., Chen C.Y., Khare K., Chen Y., Li Q., Coatings, 8, 2018, 70. https://doi.org/10.3390/coatings8020070

[83] Feng L., Wang J., Shi X., Chai C., Applied Physics A, 125, 2019, 261. https://doi.org/10.1007/s00339-019-2562-4

[84] Feng L., Zhao L., Qiang X., Liu Y., Sun Z., Wang B., Applied Physics A, 119, 2015, 75-83. https://doi.org/10.1007/s00339-014-8959-1

[85] Haryono M.B., Lin K.W.Y., Thant K.K.S., Subannajui K., Waritanant T., Journal of Physics - Conference Series, 2696, 2024, 012001. https://doi.org/10.1088/1742-6596/2696/1/012001

[86] Kan T., Xu J., Xie J., Journal of Physics - Conference Series, 2230, 2022, 012027. https://doi.org/10.1088/1742-6596/2230/1/012027

[87] Rezayi T., Entezari M.H., Surface and Coatings Technology, 309, 2017, 795-804. https://doi.org/10.1016/j.surfcoat.2016.10.083

[88] Li K., Zeng X., Li H., Lai X., Applied Surface Science, 346, 2015, 458-463. https://doi.org/10.1016/j.apsusc.2015.03.130

[89] Zhou W., Yang F., Yuan L., Diao Y., Jiang O., Pu Y., Zhang Y., Zhao Y., Wang D., Materials,15, 2022, 8634. https://doi.org/10.3390/ma15238634

[90] Song H.J., Shen X.Q., Ji H.Y., Jing X.J., Applied Physics A, 99, 2010, 685-689. https://doi.org/10.1007/s00339-010-5593-4

[91] Wang J., Chen H., Materials Express, 10[8] 2020, 1346-1351. https://doi.org/10.1166/mex.2020.1739

[92] Jiang W., Mao M., Qiu W., Zhu Y., Liang B., Industrial and Engineering Chemical Research, 56, 2017, 907-919. https://doi.org/10.1021/acs.iecr.6b03936

[93] Cheirmakani B.M., Karthikeyan M., Balamurugan S., Robert R.B.J., Results in Surfaces and Interfaces, 14, 2024, 100200. https://doi.org/10.1016/j.rsurfi.2024.100200

[94] Zhang R., Liu J., Li Z., Cui G., Journal of Physics - Conference Series, 2383, 2022, 012130. https://doi.org/10.1088/1742-6596/2383/1/012130

[95] Fan K., Jin Z., Bao Y., Wang Q., Niu L., Sun J., Song J., Colloids and Surfaces A, 590, 2020, 124495. https://doi.org/10.1016/j.colsurfa.2020.124495

[96] Jiang W., Liu Y., Wang J., Li R., Liu X., Ai S., Coatings, 12, 2022, 737. https://doi.org/10.3390/coatings12060737

[97] Varshney P., Mohapatra S.S., Kumar A., Journal of Bio- and Tribo-Corrosion, 7, 2021, 76. https://doi.org/10.1007/s40735-021-00518-3

[98] Wang P., Yao T., Sun B., Ci T., Fan X., Han H., RSC Advances, 7, 2017, 39699. https://doi.org/10.1039/C7RA06836K

[99] Satyarathi J., Kumar V., Kango S., Sharma N., Verma R., IOP Conference Series - Materials Science and Engineering, 1248, 2022, 012015. https://doi.org/10.1088/1757-899X/1248/1/012015

[100] Zhang H., Yang J., Chen B., Liu C., Zhang M., Li C., Applied Surface Science, 359, 2015, 905-910. https://doi.org/10.1016/j.apsusc.2015.10.191

[101] Fan Y., He Y., Luo P., Chen X., Liu B., Applied Surface Science, 368, 2016, 435-442. https://doi.org/10.1016/j.apsusc.2016.01.252

[102] Fan Y., He Y., Luo P., Chen X., Liu B., Applied Surface Science, 368, 2016, 435-442. https://doi.org/10.1016/j.apsusc.2016.01.252

[103] Cui S., Zhai H.M., Li W.S., Tong W., Li X.S., Cai A.H., Fan X.J., Li X.Q., Xiong D.S., Rare Metals, 42[2] 2023, 629-644. https://doi.org/10.1007/s12598-022-02130-x

[104] Du C., He X., Tian F., Bai X., Yuan C., Coatings, 9, 2019, 398. https://doi.org/10.3390/coatings9060398

[105] Latthe S.S., Sudhagar P., Devadoss A., Kumar A.M., Liu S., Terashima C., Nakata K., Fujishima, A., Journal of Materials Chemistry A, 3, 2015, 14263-14271. https://doi.org/10.1039/C5TA02604K

[106] Gao X., Guo Z., Journal of Colloid and Interface Science, 512, 2018, 239-248. https://doi.org/10.1016/j.jcis.2017.10.061

[107] Vigdorovich V.J., Tsygankova L.E., Uryadnikova M.N., Emelyanenko K.A., Chulkova E.V., Uryadnikov A.A., International Journal of Corrosion and Scale Inhibition, 10[3] 2021, 1157-1167.

[108] Almufarij R.S., El Sayed H.A.F., Mohamed M.E., Materials, 16, 2023, 4728. https://doi.org/10.3390/ma16134728

[109] Han Y., Liu Z., Pan W., Sun J., Journal of Physics - Conference Series, 1948, 2021, 012212. https://doi.org/10.1088/1742-6596/1948/1/012212

[110] Rafieazad M., Jaffer J.A., Cui C., Duan X., Nasiri A., Materials, 11, 2018, 1577. https://doi.org/10.3390/ma11091577

[111] Kim J.H., Mirzaei A., Kim H.W., Kim S.S., Applied Surface Science, 439, 2018, 598-604. https://doi.org/10.1016/j.apsusc.2017.12.211

[112] Lu Y., Guan Y.C., Li Y., Yang L.J., Wang M.L., Wang Y., Colloids and Surfaces A, 604, 2020, 125259. https://doi.org/10.1016/j.colsurfa.2020.125259

[113] Li H., Yu S., Han X., Zhao Y., Colloids and Surfaces A, 503, 2016, 43-52. https://doi.org/10.1016/j.colsurfa.2016.05.029

[114] Wang Y., Wang W., Zhong L., Wang J., Jiang Q., Guo X., Applied Surface Science, 256[12] 2010, 3837-3840. https://doi.org/10.1016/j.apsusc.2010.01.037

[115] Liu Q., Chen D., Kang Z., ACS Applied Materials and Interfaces, 7[3] 2015, 1859-1867. https://doi.org/10.1021/am507586u

[116] Zhang X., Shen J., Hu D., Duan B., Wang C., Surface and Coatings Technology, 334, 2018, 90-97. https://doi.org/10.1016/j.surfcoat.2017.11.029

[117] Li W., Kang Z., Surface and Coatings Technology, 253, 2014, 205-213. https://doi.org/10.1016/j.surfcoat.2014.05.038

[118] Zheng T., Hu Y., Pan F., Zhang Y., Tang A., Journal of Magnesium and Alloys, 7[2] 2019, 193-202. https://doi.org/10.1016/j.jma.2019.05.006

[119] Zang D., Zhu R., Wu C., Yu X., Zhang Y., Scripta Materialia, 69[8] 2013, 614-617. https://doi.org/10.1016/j.scriptamat.2013.07.014

[120] Ishizaki T., Shimada Y., Tsunakawa M., Lee H., Yokomizo T., Hisada S., Nakamura K., ACS Omega, 2, 2017, 7904-7915. https://doi.org/10.1021/acsomega.7b01256

[121] Li J., Liu Q., Wang Y., Chen R., Takahashi K., Li R., Liu L., Wang J., Journal of the Electrochemical Society, 163[5] 2016, C213-C220. https://doi.org/10.1149/2.0801605jes

[122] Li Q., Zhang X., Ben S., Zhao Z., Ning Y., Liu K., Jiang L., Nano Research, 16[2] 2023, 3312-3319. https://doi.org/10.1007/s12274-022-4937-7

[123] Yuan J., Wang J., Zhang K., Hu W., RSC Advances, 7, 2017, 28909. https://doi.org/10.1039/C7RA04387B

[124] Feng L., Zhu Y., Fan W., Wang Y., Qiang X., Liu Y., Applied Physics A, 120, 2015, 561-570. https://doi.org/10.1007/s00339-015-9215-z

[125] She Z., Li Q., Wang Z., Li L., Chen F., Zhou J., Chemical Engineering Journal, 228, 2013, 415-424. https://doi.org/10.1016/j.cej.2013.05.017

[126] Zhu J., Jia H., Materials, 13, 2020, 4007. https://doi.org/10.3390/ma13184007

[127] Emelyanenko K.A., Chulkova E.V., Semiletov A.M., Domantovsky Z.G., Palacheva V.V., Emelyanenko A.M., Boinovich L.B., Coatings, 12, 2022, 74. https://doi.org/10.3390/coatings12010074

[128] Zhu J., Bai Z., Dai X.J., Sun B., AIP Advances, 8, 2018, 075125. https://doi.org/10.1063/1.5031905

[129] Mohamed M.E., Abd-El-Nabey B.A., ECS Journal of Solid State Science and Technology, 9, 2020, 061006. https://doi.org/10.1149/2162-8777/ab9dc7

[130] Mirzadeh M., Dehghani K., Rezaei M., Mahidashti Z., Colloids and Surfaces A, 583, 2019, 123971. https://doi.org/10.1016/j.colsurfa.2019.123971

[131] Sun Y., Wang L., Gao Y., Guo D., Applied Surface Science, 324, 2015, 825-830. https://doi.org/10.1016/j.apsusc.2014.11.047

[132] Tsai B.F., Chen Y.C., Ou S.F., Wang K.K., Hsu Y.C., Internatioal Journal of Applied Ceramics Technology, 16[1] 2019, 211-220. https://doi.org/10.1111/ijac.13104

[133] Abbasi S., Nouri M., Rouhaghdam A.S., Thin Solid Films, 762, 2022, 139541. https://doi.org/10.1016/j.tsf.2022.139541

[134] Gao Y., Sun Y., Guo D., Applied Surface Science, 314, 2014, 754-759. https://doi.org/10.1016/j.apsusc.2014.07.059

[135] Li S.Y., Li Y., Wang J., Nan Y.G., Ma B.H., Liu Z.L., Gu J.X., Chemical Engineering Journal, 290, 2016, 82-90. https://doi.org/10.1016/j.cej.2016.01.014

[136] Xiang G.X., Li S.Y., Song H., Nan Y.G., Microelectronic Engineering, 233, 2020, 111430. https://doi.org/10.1016/j.mee.2020.111430

[137] Li J., Xu J., Lian Z., Yu Z., Yu H., Optics and Laser Technology, 126, 2020, 106129. https://doi.org/10.1016/j.optlastec.2020.106129

[138] Hu L., Zhang L., Wang D., Lin X., Chen Y., Colloids and Surfaces A, 555, 2018, 515-524. https://doi.org/10.1016/j.colsurfa.2018.07.029

[139] Zhang Y., Chen G., Wang Y., Zou Y., Surface Review and Letters, 28[5] 2021, 2150027. https://doi.org/10.1142/S0218625X2150027X

[140] Zhu M., Tang W., Huang L., Zhang D., Du C., Yu G., Chen M., Chowwanonthapunya T., Materials, 10, 2017, 628. https://doi.org/10.3390/ma10060628

[141] Fadeeva E., Truong V.K., Stiesch M., Chichkov B.N., Crawford R.J., Wang J., Ivanova E.P., Langmuir, 27[6] 2011, 3012-3019. https://doi.org/10.1021/la104607g

[142] Manoj T.P., Rasitha T.P., Vanithakumari S.C., Anandkumar B., George R.P., Philip J., Applied Surface Science, 512, 2020, 145636. https://doi.org/10.1016/j.apsusc.2020.145636

[143] Rasitha T.P., Thinaharan C., Vanithakumari S.C., Philip J., Colloids and Surfaces A, 636, 2022, 128110. https://doi.org/10.1016/j.colsurfa.2021.128110

[144] Rasitha T.P., Philip J., Applied Surface Science, 585, 2022, 152628. https://doi.org/10.1016/j.apsusc.2022.152628

[145] Gao X., Tong W., Ouyang X., Wang X., RSC Advances, 5, 2015, 84666-84672. https://doi.org/10.1039/C5RA15293C

[146] Zhang X., Wan Y., Ren B., Wang H., Yu M., Liu A., Liu Z., Micromachines, 11, 2020, 316. https://doi.org/10.3390/mi11030316

[147] Yang Z., Zhu C., Zheng N., Le D., Zhou J., Materials, 11, 2018, 2210. https://doi.org/10.3390/ma11112210

[148] Wang Y., Chen J., Yang Y., Liu Z., Wang H., He Z., Nanomaterials 2022, 12, 2086. https://doi.org/10.3390/nano12122086

[149] Wang L., Li H., Song J., Sun Y., Surface and Coatings Technology, 302, 2016, 507-514. https://doi.org/10.1016/j.surfcoat.2016.06.057

[150] Long D.P., Xue J.R., Yan Z.X., Advanced Materials Research, 834-836, 2014, 29-32. https://doi.org/10.4028/www.scientific.net/AMR.834-836.29

[151] Shen Y., Tao J., Tao H., Chen S., Pana L., Wang T., Soft Matter, 11, 2015, 3806-3811. https://doi.org/10.1039/C5SM00024F

[152] Shen Y., Tao J., Tao H., Chen S., Pana L., Wang T., RSC Advances, 5, 2015, 32813-32818. https://doi.org/10.1039/C5RA01365H

[153] He H., Qu N., Zeng Y., Surface and Coatings Technology, 307[A] 2016, 898-907. https://doi.org/10.1016/j.surfcoat.2016.10.033

[154] Kuzina E.A., Omran F.S., Emelyanenko, Boinovich L.B, , Colloid Journal, 85[1] 2023, 59-65. https://doi.org/10.1134/S1061933X22600567

[155] Han J., Cai M., Lin Y., Liu W., Luo X., Zhang H., Wang K., Zhong M., RSC Advances, 8, 2018, 6733. https://doi.org/10.1039/C7RA13496G

[156] Kuzina E.A., Emelyanenko K.A., Teplonogova M.A., Emelyanenko A.M., Boinovich L.B., Materials, 16, 2023, 196. https://doi.org/10.3390/ma16010196

Milton Keynes UK
Ingram Content Group UK Ltd.
UKHW021133040824
446423UK00008B/17